ABOUT THE AUTHOR

Peter Wohlleben spent over twenty years working for the forestry commission in Germany before leaving to put his ideas about ecology into practice. He now runs an environmentally-friendly woodland as well as caring for both wild and domestic animals.

He is celebrated for his distinctive ability to bring together ground-breaking scientific research with his own observations of nature. The author of many books, they include *The Inner Life of Animals* and the international bestseller *The Hidden Life of Trees*.

By the same author:

The Hidden Life of Trees

The Inner Life of Animals

The Secret Network of Nature

PETER WOHLLEBEN

THE WEATHER DETECTIVE

Rediscovering Nature's Secret Signs

1 3 5 7 9 10 8 6 4 2

Rider, an imprint of Ebury Publishing,
20 Vauxhall Bridge Road,
London SW1V 2SA

Rider is part of the Penguin Random House group of companies whose
addresses can be found at global.penguinrandomhouse.com

Penguin
Random House
UK

Copyright © pala-verlag, Darmstadt, 2012

Peter Wohlleben has asserted his right to be identified as the author of this
Work in accordance with the Copyright, Designs and Patents Act 1988

Translation copyright © Ruth Ahmedzai Kemp 2018

Peter Wohlleben has asserted her right to be identified as the author of this
Work in accordance with the Copyright, Designs and Patents Act 1988

First published in Germany as *Kranichflug und Blumenuhr* in 2012
First published by Rider in 2018
This edition published by Rider in 2019

www.penguin.co.uk

A CIP catalogue record for this book is available from the British Library

ISBN 9781846046025

Typeset in 11.75/14.75 pt Minion Pro
by Integra Software Services Pvt. Ltd, Pondicherry

Printed and bound in Great Britain by Clays Ltd, Elcograf S.p.A.

Penguin Random House is committed to a sustainable future for
our business, our readers and our planet. This book is made
from Forest Stewardship Council® certified paper.

CONTENTS

INTRODUCTION:
on Nature's Trail

THE moment we step out of the door and stroll through the garden or a nearby park, we are surrounded by nature. Thousands of processes, from the minute to the gargantuan, are unfolding all around us, and they are fascinating and beautiful to behold – if only we open up our senses and take notice of them.

In the past, it was vital that everyone could recognise and interpret these signs. People were dependent on nature and intimately familiar with it. Nowadays, fully stocked supermarket shelves, constant energy supplies and measures in place to insure us against any conceivable act of nature all trick us into thinking that we no longer rely on our ancient bond with the natural world. Our distance from nature is particularly obvious during hot, dry summers. While farmers and foresters are desperate for rain, most of the urban population is delighted to hear forecasts predicting ongoing dry weather, oblivious to the impact of a prolonged drought. And yet, in the face of climate change and damage to the environment, it is more urgent than ever that we recognise and understand the signs of nature. Only then will we appreciate what we stand to lose.

Television, radio and the internet all make gazing out of the window to find out what the weather is like rather redundant. We have countless specialised services at our fingertips to let us know what is going on outside in the garden. There are regular updates for everything we could possibly imagine wanting to know about – from whether we're faced with rain or shine, to when birds will migrate or aphids hatch – and such information is readily available for anyone interested to look it up. If you want even more precise prognostic data, you can simply install an electronic weather station outside that sends a live feed to you in the comfort of your living room.

But if you enjoy gardening and spending time in nature, you can manage perfectly well without these bulletins updating you constantly about the weather. We can glean most of the same information from clues around the garden, from the animals and plants in our local area, in fact even from the inanimate environment. Whether it's forecasting what's ahead or assessing current weather events, whether it's insect infestations or when it's safe to say a season has started or ended, you can read all of this data from your garden much more accurately than any newsreader from an autocue. There can be a huge difference, after all, between your garden and another location just a few miles away in terms of how a natural event unfolds and the impact it has. And that is ultimately why we look to media forecasts: to assess the situation on our doorstep.

This guide will help you to decipher the vast quantities of information you can glean from your local environment and especially your garden. You can become your own nature expert. It will address many everyday questions that in the future you'll be able to answer for yourself; and many phenomena will suddenly be easier to understand when you know the background.

The most important motivation for writing this book was the prospect of encouraging more people to take pleasure in time spent outdoors and relaxing outside. How wonderful it is to experience things consciously that you had until now passed by obliviously. How exciting it is to foresee changes in the weather, and in flora and fauna, before they happen. When we are out and about, experiencing our surroundings with all our senses, nature is closer to us than ever before. And the ancient bond between us and our environment can be renewed.

WHAT WILL THE WEATHER BE LIKE?

EVERY TV or radio news bulletin is always followed by the weather report, and it is often better than it's reputed to be. Forecasts of up to a week in advance are about 70 per cent likely to be true, while there is a 90 per cent success rate for those covering 24 hours ahead. Looking at it the other way, this means every tenth weather forecast misses the mark. The reason for this is chaotic weather conditions which simply cannot be predicted. I find it very irritating that presenters never admit to this with a statement like, 'Because of the current situation, today's data is very uncertain.' You simply never hear that. Nevertheless, it can't hurt to look outside and to read the signs for yourself if you wish to check what the clouds are up to. Over the years, you stand to develop a strong sense of what is going to unfold in the next few hours.

Cloud towers and rosy sunsets

The evening Sun is a much-loved prophet. If it sets with a warm, rosy glow, it is taken as a sign of sunshine the following morning, as in the rhyme, 'Red sky at night: shepherd's delight.' This happens because the sunbeams stream in low through the atmosphere from the clear skies in the west and light up the clouds slowly drifting off to the east. And since, in western Europe, the weather usually comes from the west, a broadly cloudless western horizon means clear skies for the following few hours.

Things are the other way around with rosy dawn skies. The saying goes, 'Red sky in the morning: shepherd's warning.' This is also usually right. For the Sun rises in the east, where the sky is still clear, and shines onto the clouds gathering in the west, which will rapidly spread and fill the sky.

Every rule has its exception, of course: when the wind blows not from the west, but from the south or the east, red skies at sunset or sunrise bear no prophetic significance.

The wind direction can itself be used as a forecasting instrument. The west wind carries moist sea air from the Atlantic, which form clouds and often rainfall. As clouds insulate the Earth like a blanket, they influence the temperature. Dense cloud cover in the winter prevents it from dropping as severely as when there are blue skies, by reducing heat loss at night. However, there is a greater chance of rain with a westerly wind. In summer, meanwhile, cloud cover prevents hotter spells, as it keeps the Earth's surface in the shade.

South winds bring warmth from the Mediterranean or even the Sahara. In summer, these southerly winds can trigger a heat wave and in winter they often carry storms in their luggage. This is because on their way across Central Europe they meet polar air masses which flow to us from the north, bringing about a violent exchange as the cold air mingles with the hot air. This can, of course, also happen with cold north winds as they come into contact with unusually warm winter air.

The east wind promises stable conditions and a clear sky. In summer, it is very warm and in winter bitterly cold. Without protective cloud cover, every season shows its extreme side.

To determine the wind direction, you can't beat the classic weathercock. The cockerel spins about on his seat of a cross, whose four arms each bear a letter for the four points of the compass. Why not install a weather vane like this in your garden or on the roof of your house? As the cock always looks in the direction from which the wind blows, assuming it is correctly installed, it shows the wind direction and thus allows you to predict the coming weather.

The key players in determining the weather, however, are the clouds. Whether the weather will be fine or poor – whatever our criteria – depends on the presence of clouds and their cargo, raindrops. If a low-pressure area emerges, the air literally becomes thinner (as in a tyre if you let some air out). The water vapour cannot dissolve completely in this thinner air and becomes visible in the form of clouds.

An early harbinger of a bad weather front is the appearance of artificial clouds, i.e. the condensation trails of aeroplanes (contrails). If these don't dissolve, that means humidity is on its way, and with it a low-pressure area. The sky will soon cloud over.

We can also generally rely on the following rule of thumb. The weather always changes when the clouds approach from a different direction to that of the wind at ground level, which can lead to the appearance of beautiful, small, fluffy clouds.

We can tell how dense the layer of cloud above us is from its colouring: thinner clouds appear white, because some sunlight is still able to travel through them. Thicker, taller clouds, on the other hand, appear grey or even black, because barely a single beam of light can battle through these immense towers of water droplets. The higher these structures, the sooner it will rain.

BUZZARDS ON THE RISE

When the Sun's rays heat the earth, the layer of air nearest the ground also warms up. This results in a temperature gradient moving upwards. Warm air has the tendency to rise as it is less dense than cold air. It doesn't do this in a uniform way, however, but it forms invisible, tubular structures with a diameter of anything between a few metres and a few hundred metres. Warm gas rises high into the atmosphere, and in turn, cold air sinks to the ground at the edges of these tubes. What we're talking about here, of course, is thermals. There is an indirect way that you can observe this fascinating phenomenon. On a fine day, you will see individual fluffy cumulus clouds forming at the top of a column of rising warm air, where it cools and condenses into water droplets.

In terms of animal behaviour, you'll know you're looking at a thermal when you see circling birds of prey. They use the upward lift to soar for hours without a single wing stroke. They can only maintain this, however, for as long as they remain within the thermal column. And as these shift (which you can see from the drift of the clouds), buzzards and kites also gradually move along with them. Migratory birds use the warm air to gain height without expending too much effort. You will often see crows suddenly start to circle for about 15 minutes until they leave the area of lift, a storey higher than they started, and continue on their way.

During longer periods of bad weather, the whole thing stops working. No Sun – no lift. One exception is on mountain slopes, beaten by rainy winds, as here again the air masses move upwards. And here you'll also find the birds that want to soar high.

Precipitation is formed by two processes. One way is that water droplets collide and form ever larger drops. This is a very slow process and the result is a long-lasting drizzle, more typical of flatter clouds. Larger rain drops can only form in higher towers of cloud, because for this process ice comes into play. At the top of the cloud it is very cold, and here the water freezes. In no time at all, more water clings to the ice crystals, immediately freezing on contact. These ice crystals become too heavy to remain airborne and fall to the ground. On the way down, they thaw as the air gets warmer and the result is very large rain drops. From this you can conclude that the larger the droplets, the thicker the clouds must be, and the greater the quantity of rainfall per minute.

Every heavy raindrop was once an ice crystal or a large snowflake. If the flake doesn't melt on its way down to earth, it will be snow. Strictly speaking, it can also snow in summer; it's just that the snow melts high up, long before it reaches us.

Speaking of snow, there is something else we can tell from a snowflake's size and consistency. Basically, the smaller the flakes, the colder the air and the greater the chance of it settling. This is because cold air can hold almost no liquid water, so the flakes don't take on more water, which is what makes them grow in size.

Meanwhile, larger snowflakes indicate warmer weather. They keep accumulating water vapour and growing bigger and bigger until shortly before they reach us. Sometimes it snows great clumps of snowflakes, but their splendour is only short-lived. And because these thick flakes usually contain a lot of moisture, this seemingly harmless kind of snowfall in fact brings with it considerable hazards. Landing on branches or power lines, the snow forms into a thick layer that gathers for some time without falling. The accumulated weight of this 'damp snow' can cause branches and pylons to collapse, as well as the entire roofs of buildings.

Snowmen can also be used for weather forecasting. It is only in relatively mild weather that snow has the right consistency to be rolled into balls. Therefore, building a snowman can also mean spring is around the corner, unless another cold spell follows, of course.

But back to the clouds. If you see tall, towering clouds on the horizon, it means rain (or snow) is likely soon. If they seem to puff out at the top, or form an anvil shape (where the cloud tower is being pulled apart at the top), then a thunderstorm is on the way. Shortly before the storm front vents its fury, the wind grows brisk and strong, perhaps even reaching hurricane levels. It is only when the heavens open that the wind drops again, almost instantaneously.

After the rain front passes, it usually gets colder. This is because a low-pressure area (which brings the rain with it) is drawn across the country with a warm front, and what follows is a cold front. Both fronts spell rain, but in between it often clears up briefly. Until the cold front passes by completely, however, the sunny intervals do not signify a shift to better weather. Short showers will continue until the low pressure finally moves on.

A special case is fog and its by-products: dew and hoarfrost. It becomes foggy when the water vapour can no longer disperse into the air because the air is already saturated. Cold air can't hold much water; unlike warm air, which can hold a lot. This is why foggy weather is particularly common in the colder half of the year, while in summer there is usually good visibility. Incidentally, a hair-dryer works according to precisely this principle: the air around the hair is heated so that it can absorb more water, and the hair dries.

If the temperature drops sharply at night, the air can no longer hold the water and 'sweats' it out. Small droplets accumulate on the ground as dew or, if the air temperature falls below freezing, as hoarfrost. When you see this phenomenon, which is combined with a drop in temperature, in the garden in the morning, or on the roof-tiles of the house next door, you can generally bet that the weather that day will be fine. Such a sharp drop in temperature is caused by relatively dry air, with little excess water to form clouds. Without the cosy blanket of cloud cover, the landscape cools down sharply.

Plants as weather prophets

When a high-pressure fine spell subsides and a low-pressure system sits threateningly on the doorstep, the air humidity gradually rises. And many plants don't like this, because the coming rain plays havoc with their offspring. Many species send their seeds off on their way borne on small fluffy hairs, which are carried away by even the gentlest breeze. But when they're wet, these little hairs are effectively grounded; a rain shower flushes all the splendour from the blossom down to the ground beneath the mother plant. The opportunity to conquer new territories is lost, squandered.

The same applies to the pollen in fresh flowers: if knocked to the ground by rain, it can't be couriered away by bees and used for fertilisation. When the air gets more humid, suggesting rain is on the way, certain flowers react with a precautionary measure, closing their petals protectively over their interior. One example is the silver thistle, now a protected species. Its large flowers are particularly decorative and the way it folds up is no less striking. It's not for nothing that their common name in German is 'weather thistles'. The forecasting works even with dried plants, since it is based on a purely mechanical process. The outer petals swell with a rise in air humidity and stand up on end. In the past, people used to hang these flowers by their front door to give early warning of impending rain.

There are indeed other plants whose flowers react to changes in the weather, such as the gentian or the water lily. In the case of aquatic plants, the ability to react to a change in moisture makes little sense: water lilies, for example, sit in water the whole time anyway. And yet their blossoms are nevertheless a reliable indicator of a coming change of weather. It is not yet clear whether the trigger is the pressure difference (high or low pressure) or only the diminishing brightness of a cloudy sky. But it certainly seems to be a reliable forecaster. The flowers close when they sense rain, often hours before it comes.

I would like to highlight one more example: the daisy. It grows practically everywhere, and if you don't already have some in your garden, I would certainly reserve a corner for them. One glance at the white and yellow flowers is enough to tell if you should hang your laundry out in the garden or if inside would be safer. If rain is on the way, or a storm, the petals close up. Some also droop downwards, to avoid letting a single drop in. When the weather is fine, the blossom remains open. The entire response mechanism only functions during the day, however, because daisies always close up shop in the evenings, like many other flowers.

In the case of daisies, this opening and closing mechanism is well understood; it's a matter of thermonastic motion. This term refers to the difference in growth between the upper and lower sides of a petal. The upper side grows faster at higher temperatures than the lower side. In the warm sunshine, the flower therefore opens up, while dark rain clouds cause cooling temperatures, encouraging the underside to grow faster, and making the petals close up. This process explains why they close up at night, when it's cooler. For the daisy to be able to react at any moment, the petals need to be constantly growing, and so they grow longer day by day, bit by bit. This means you can also distinguish younger flowers from older ones.

However, even among the colourful weather prophets, not every bloom takes part in this to and fro process. There are flowers that leave their pollen and nectar open to takers even in the rain. Some cultivated varieties might perhaps have lost this ability to react. Or maybe some of these lone wolves want to make themselves attractive to less rain-shy insects and to gain an advantage in the pollination stakes. There remain so many mysteries to be explained.

Animal weather prophets

Besides animals that react to the rain itself, there are those that demonstrate behavioural changes in advance of a rainstorm's arrival. One species is particularly renowned for this meteorological quality: the minuscule thrip, an insect just one to two millimetres in length, which is also known variously by the common names thunderfly, thunderbug and storm fly. Thrips have fringed wings, although these are really more like paddles, which the tiny insects use to propel themselves through the air. For creatures of this size, air has the same resistance as water does for us humans, giving them a certain buoyancy. The result is that these diminutive beasts don't fly in the true sense of the word. Their motion is more like swimming through the air, and it's therefore a rather slow action. The conditions they love best are when it's hot and sticky and when there's good air movement; with a warm breeze, they can travel from plant to plant much more efficiently. It is precisely these conditions (sultry air and rising winds) that emerge in the run-up to a storm, so that is when you'll see the air swarm with these microscopic pests. Seeing them can therefore be an early warning of an impending storm.

Swallows, on the other hand, are less reliable as weather prophets than convention dictates. It is said that low-flying swallows mean the heavens are about to open. The reason is the abundance of insects flying low over the grass. But researchers have discovered that, if anything, it is the other way round: as the wind picks up before a storm, swallows are likely to fly higher than usual. The saying 'when swallows fly high, the weather will be dry' could well lead you astray.

The chaffinch has its own idiosyncratic way of warning us of a change in the weather: it modifies its song when the weather is set to turn. The males usually trill a melody that sounds a little like 'chi-chip-chirichirichiri-chip-cheweeoo'. (At my forestry college in Germany, we were taught a word pattern to remember their call – *Bin bin bin ich nicht ein schöner Feldmarschall*? ('Aren't I, aren't I, aren't I a lovely field marshal?') – and their song does usually fit this mnemonic phrase, while the Sun is shining, at least.) But this sunny call is reserved for fine weather. If storm clouds loom or it starts to rain, the chaffinch switches to a rather monosyllabic chant. His so-called 'rain call' is a simple 'raaatch'. Here too experts disagree on whether or not the male chaffinch's behaviour can be relied on for forecasting. He clearly reserves his 'raaatch' call for disturbances of various kinds, not just heavy rain. I spend a lot of time in old deciduous forests which are teeming with chaffinches. When I show up, my presence is a (minor) disturbance, and yet they continue to chirp away undisturbed, trilling their 'sunshine call'. It is only when the weather changes that I hear them adopt their rain call. But judge for yourself how reliably your native finches moderate their song according to the changing circumstances.

And what about you?

It is not at all unusual for people too to sense physical changes in the weather. After all, high and low pressure areas are so called precisely because the air pressure differs considerably in them. If a high is followed by a low, it's like letting some air out of a tyre. The device that measures this pressure is called a barometer. It functions in the same way as the tyre pressure gauge at a petrol station. And for us, living in the Earth's atmosphere, it's just like sitting inside a gigantic car tyre.

Some people have a kind of built-in barometer, suffering pain or discomfort when the air pressure drops. The term for this is 'weather sensitivity' (or meteorosensitivity), but there is not yet consensus about it in the scientific community. One theory suggests there's a change in the conductivity of cell membranes in the body. The sensitivity threshold of the nervous system is lowered, so that pain occurs more easily. People who have an acute medical condition seem to be particularly affected.

Other experts attribute the symptoms to changes in the air mass; that is, to the rapid change from warm, dry air to cold, damp air. Much remains unexplained, but one thing is clear: for some people, inclement weather makes itself known through physical complaints. Pay attention to what happens to you next time the barometer shows a particularly strong drop in pressure. Perhaps you'll find you can dispense with this measuring device, after all.

2

IS IT WINDY
OR COLD?

Our Earth is surrounded by a delicate layer of gases: the atmosphere. This separates us from the universe and is, depending on the definition, about 100 km thick. Or should we say thin? Because the density of the atmosphere decreases rapidly with altitude, making the air too thin for us to breathe at an altitude of just a few kilometres.

This delicate structure shelters us from cosmic rays. Distant stars and other celestial bodies, but above all our Sun, pelt a ceaseless shower of protons and atomic nuclei down at the Earth. We wouldn't last long if fully exposed to it, but thankfully the atmosphere filters out most of this lethal radiation. Our air cushion also balances out the immense temperature differences between day and night. Our companion in the universe, the Moon, makes for a telling comparison: it has no atmosphere and therefore no buffer. Night temperatures on the barren crater landscape plummet to an icy −160 °C, while during the day the thermometer climbs to a staggering 130 °C.

The Earth's air bubble contains 21 per cent oxygen, a very aggressive gas. We perhaps shouldn't take it as a given that it is necessary for life: there was a time after all when there was no oxygen, only water vapour and carbon dioxide to 'breathe'. Higher life forms did not yet exist, and the prevailing cyanobacteria got by just fine in this primordial atmosphere. At least they did until they had polluted all the air with the gas they exhaled: oxygen. Bad for the bacteria, but good for the other life forms, which now adapted to the new conditions and threw themselves into a steady uphill development. This is all 2.4 billion years in the past, but in hidden corners around our globe, for example at the bottom of the oceans, certain bacterial species still survive that breathe hydrogen and sulphur instead of oxygen.

You can see evidence of oxygen's aggressive nature every day in the garden. It attacks the metal parts of your hoes and spades, resulting in rust. And because iron is also found in many rocks, rust formation takes place there as well, and is responsible for the reddish colouring of stones and sand in certain areas.

The air is also an important transport route for animals and plants. Seeds fly with the wind to new locations, and for birds, our atmospheric bubble provides an ecological niche, as useful for long-distance journeys as it is for hunting insects. Some species spend almost their entire lives in the air and come to the ground only to nest. For example, the common swift sometimes flies for several months without interruption, even sleeping on the wing, albeit for a mere few seconds at a time.

The air plays a fundamental role with respect to something else: the weather. Due to the temperature differences between the poles and the equator, there is a constant exchange of warm and cold air around the planet. The rotation of the Earth causes further deflection and acceleration, so that the air masses are constantly on the move. These masses carry water vapour with them, which rises up from the seas and forests and is offloaded again many thousands of miles away in the form of precipitation. In Europe, we owe our favourable climate and constant supply of water to the oceans. Areas of rain migrate to us from the Atlantic in the west. The evaporated seawater brings lush life to fields and forests, replenishing rivers and lakes. This transport of water vapour only works with moving air masses, i.e. the winds.

Measuring wind speeds

I'm often astonished by the technological prowess we have at our fingertips these days when it comes to domestic meteorological devices. When I go to the department store, I stare in wonder at the shelves of electronic home weather stations, which can even measure the wind speed and transmit the data wirelessly to my home. Is there really any need nowadays to step out of the house to know what conditions the garden is currently exposed to? Well, these devices may give a reliable reading of air temperature, but with rainfall and wind speed their accuracy is limited. After all, the unit can only detect what is happening in one precise location. In another corner of the garden, only 10 or 20 metres away, the conditions might be quite different. In the case of storms, in particular, there can be significant differences over very short distances because storms are often accompanied by turbulence. These whirlwinds dance through the landscape in such a way that everything in their narrow path is shaken violently, yet with hardly any impact to the left and right of them. A storm front like this swept through the reserve I manage last July. A small tornado rose up, flattened a hectare of woodland, and then vanished as quickly as it came. Such tornadoes are very rare, but their smaller sibling – air turbulence is an everyday meteorological phenomenon for the garden owner.

It is therefore difficult to measure peak wind speeds with a domestic wind meter. It is much more effective to judge by taking an inventory of your garden. Trees are ideal, as are pot plants and sun screens. These are all objects you can use to establish your own wind scale.

The internationally recognised Beaufort Scale, which is outlined on many websites, categorises wind speed by 'wind force' ratings, defined by the impact of the wind on everyday objects and items in the garden. For example, at wind force 6 (strong breeze, 25–31 mph), the Scale describes it as being difficult to use an umbrella. You can refer to these descriptions to help you quite accurately determine the wind speed where you are.

The information broadcast in the daily weather forecast, on the other hand, cannot be transferred easily to your local circumstances, since it applies to large areas. The wind speed outside your back door is affected by the degree to which your property is exposed or sheltered; for instance, whether it is situated in the shade of a group of trees or other houses. It is worth making your own observations so that you are better equipped to interpret the weather forecast. For example, imagine your garden is sheltered in the lee of a hill. Using the Beaufort Scale, as described above, you will find that the shade of the hill reduces the force of each storm by one or two categories compared to the wind force forecast in the weather report. Once you establish this correlation, you can relate it to any weather forecast, and you can more reliably estimate the risk of storm damage for your garden.

The following rule of thumb is worth remembering. At 6 on the Scale, you should secure garden furniture and potted plants, if they are not already fixed firmly in place. Sun shades and umbrellas should be folded up. At 8 on the Beaufort Scale, it becomes dangerous to stand under a tree, because dead branches can break off. I would postpone a walk in the woods, but in an open field, it can be exciting to experience this elemental force of nature. From wind force 10 on the Scale, gusts of wind can blow at up to 60 mph and you would do better to stay inside, as spruces especially as well as other tree species with weak root bases (e.g. poplars) can be uprooted. Not only can this endanger you directly, but it can also wreak havoc on the roads and cause serious traffic delays.

To assess storm damage in the forest I manage, I have my own personal calibration scale. After a gale has swept through (but not while it is happening!), I always look along a forest path that runs over a particularly exposed ridge, to see if any firs have fallen. If nothing has happened there, the wind force was within the harmless range. However, if there are trees lying across the path, I would expect to find similar obstacles on the path elsewhere in the forest, so I have to check the whole territory for damage. Hikes are postponed in the area until all fallen and leaning trunks have been made safe.

Despite all the forecasts, uncertainties always remain. The decisive factor, after all, is not the average range of risk, but the strongest gust of wind that sweeps through your garden. And an individual gust can be much more violent than the forecast average. If you are going to be out of the house all day, it is advisable to take precautionary measures for your property even with forecasts of gentler winds.

Ideal temperatures and living thermometers

The air temperature needs to be within a certain range for us to survive. Worldwide, the extremes are 70°C (in south-eastern Iran) and minus 72°C (in Siberia). As of July 2017, the record temperatures in the UK are 38.5°C in Faversham, Kent and minus 27.2°C in Braemar, Aberdeenshire. We tend to feel most comfortable, though, at around 21°C with dry air. This preference is linked to our evolutionary past in the African savannah, the climatic zone that was the cradle of humankind.

To measure the air temperature, all we need to do nowadays is look at the thermometer on the wall, and we immediately know what it's like inside and out. However, these values don't tell us anything about how we will feel, since the sensation of heat depends on several other internal and external factors, for example humidity. The more water vapour the air contains, the colder you feel at the same temperature, because water is a better conductor than air and therefore your body heat dissipates faster. I tend to look only at the thermostat display when I feel that it's too cold or too warm in the house, meaning that the thermometer serves only as a confirmation of my assessment that the heating ought to go on or off.

Animals and plants don't have these aids, but they are also completely superfluous for them (as they are for me). Just as with us humans, every species has temperature sensors that determine their behaviour. And this behaviour allows us to draw conclusions about the outside temperature.

Let's start with our biological thermometer at the bottom of the scale. To identify temperatures below zero degrees we need no further clues, since a glance at the frozen puddles or water butt tells us all we need to know.

There's an insect that can help us work out if the temperature is just above freezing point: the winter crane fly, also known as a winter midge. We still know very little about how they live. Despite resembling an oversized mosquito, we know at least that they don't bite us humans. They have a kind of biological antifreeze in their blood, so that they are not affected by the cold. The dark colouring of its chitin exoskeleton and wings helps the insect warm up rapidly even in the weak winter Sun, so they can be seen in small swarms buzzing above your flowerbeds in February even at temperatures just over zero degrees.

When it is just a little warmer, from about 5°C, then it's time for migratory toads and frogs to emerge. On damp spring and autumn nights with temperatures as low as this, you can reckon on encountering these migrants on the roads.

NATURE'S ANTIFREEZE

When the temperature drops below freezing, this spells
trouble for many animals. Mammals and birds keep their body heat
at a constant temperature by means of their internal combustion
processes, which uses up a lot of energy. For this reason, they will
either need to build up a decent layer of fat in autumn or find enough
food over the winter. And because many young animals in particular
won't manage either, winter is also the time of the harshest selection
process in nature. Some species avoid the risk of freezing to death by
hibernating through the winter and lowering the body's thermostat to
a few degrees above zero.

Cold-blooded creatures such as amphibians or insects don't have
this option. Their body temperature sinks mercilessly, in line with
the ambient air. When the temperature reaches zero, their cells
would simply burst if it weren't for the protective measures they have
at their disposal. There are various strategies for preventing their
cells from freezing. One approach relies on the diminutive size of
many organisms. When it's freezing, ice forms in water, crystallising
around minute particles such as dust. The smaller the volume of
fluid, the fewer particles, or nuclei, there are within it. Because of this
principle, tiny aphids can survive temperatures of less than 20 °C,
without employing any particular antifreeze in their blood. Larger
representatives of the insect kingdom, such as ladybirds and flies,
need to come up with another solution. They drain their guts to
reduce as much as possible the quantity of nuclei around which ice
could crystallise, but they cannot completely purify their body water.
This is why they produce glycerol, a substance which significantly
lowers the freezing point of the body's liquids. Larger animals such
as European amphibians need to find warmer places to hide, deep
within the earth or in deep water to survive the winter, as even with
their frost protection measures they would freeze.

A very different approach is taken by our honey bees. The
whole population sits together in what is known as a winter cluster,
maintaining a core temperature of 25 °C. This also explains why these
insects produce so much honey: they consume an enormous amount
of energy to keep the collective winter heating going.

In recent years, something has puzzled me. There's a bright blue sky in April and the fruit trees are blossoming, and yet there are no bees to be seen. The only ones out making the most of the enormous array of flowers on offer are the bumblebees, which set off to collect nectar as soon as the temperature rises above 9 °C. The young apple trees promised a rich harvest last spring, but I was seriously worried about the lack of pollinators. I quickly bought two bee populations to help out. Later (now as a more informed beekeeper) I learned that bees only leave the hive at temperatures above 12 °C. So, I should have simply waited for warmer weather; then wild bees or honeybees would have flown in from nearby.

This temperature appears to be a magic threshold for several species. For example, at the height of summer, billowing meadows full of grasses and herbs are the habitat for grasshoppers and crickets, which provide an orchestral backdrop with their chirping, known as stridulating. However, this soundscape is by no means constant. Because to really make a decent sound, the air temperature must be at least 12 °C. If it is cooler, you'll barely hear a squeak from these tiny musicians. As cold-blooded creatures, grasshoppers can't regulate their own body temperature, and only really get going when it's warm enough. Their body movements become faster with rising temperatures, resulting in ever more rapid vibrations of the legs and wings, which produce the chirping sound, depending on the type. This also changes the frequency of the tone produced: the warmer it is, the higher the pitch.

When it gets warmer outside than the optimal body temperature of 35 °C, the bees stay at home. When flying, they generate additional warmth so that they can quickly overheat. Thus, here we also have an indicator of very high temperatures.

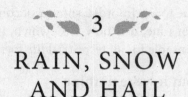

3

RAIN, SNOW
AND HAIL

It's not for nothing that the Earth is called the Blue Planet. When astronauts started photographing it from space, it appeared in the images as a blue sphere, streaked with white. The land masses faded to a few brownish patches here and there.

This was no secret, of course, but our new perspective really highlighted the extent to which the Earth is a watery planet. Over 70 per cent of the surface is covered by seas and oceans, so that the continents, which make up the remaining 29 per cent, are actually islands.

The life-sustaining water that floods our planet is likely to have originated from comets, which travelled through the universe like dirty snowballs and, veering off course, crashed into the Earth.

Liquid water is a prerequisite for all living things. As ice or as water vapour it would be hostile to life. And since there is only a very narrow temperature band (astronomically speaking) between freezing point and boiling point, we have been lucky that Earth stays at the requisite distance from the Sun.

It is not necessarily the case, however, that life on other planets is only possible under precisely these conditions. It is quite conceivable that in other places in space, a different liquid (and therefore also a different temperature range) plays the role that water plays on our planet.

However, due to a primordial shower of comets way back in the depths of time, it is now H_2O which, in the form of rain, is indispensable for all terrestrial life forms.

Rain – without it we're nothing

Water is the garden's elixir of life. Plants can dispense with soil if they need to, but not moisture. For your garden plants, growth is very much dependent on the total annual precipitation.

Refreshing rainfall often has to travel a long way before it falls from the sky. Above the distant oceans, water vapour rises into the atmosphere in the heat of the Sun, condensing in the cooler air strata over the mainland and falling back to the ground as precipitation. Seen this way, rain is a kind of liquid sunshine.

With every shower, however, it is not only chemically pure H_2O that descends to Earth. At the same time, the shower cleanses the air, bringing down with it all kinds of suspended matter, such as pollen, dust or even acid particles, to enrich the soil with nutrients (or pollutants).

You'll notice this if you consider the visibility. You can see much more clearly into the distance after the rain has given the air a 'deep clean', because there's no haze from dirt particles blocking the light rays.

How much rain is enough?

In our latitudes, water is the decisive factor for a healthy garden and successful harvests. Of course, the temperature also plays a key role, but the availability of water is to a certain extent the prerequisite for all plant life.

It's mainly in winter that the ground water is refilled. If autumn and winter are rainy enough, the subterranean reservoirs can fill up to the brim. The plants which would usually help themselves to the supply are dormant over winter and abstain from drinking. Any water that can't be held in the soil sinks into the deeper layers and becomes groundwater. It's therefore worth celebrating 'bad' weather in the colder half of the year. In the warmer seasons, many plants consume more water than is supplied from above, so they rely on these ground water supplies to keep their thirst quenched.

How much rain is enough? It's not an easy question to answer. First, we must consider the climate, whether it's humid or arid. In a humid climate, more precipitation falls on average per year than evaporates again. In the case of an arid climate, the opposite is true: overall evaporation exceeds precipitation. There are transitional climates which cross the borderline at times from a humid climate to a dry, arid one. Fortunately, the UK and most of Central Europe generally lie within the humid zone, although certain landscapes slip for some months of the year into the arid category.

So, we basically get more rain here than can evaporate again. But that isn't enough, because we might well ask ourselves at this point how much of the precipitation can be stored in the upper layers of soil. After all, it is only there that it is of any use to the plants as a resource to be called on in summer. In Chapter 8, 'Assessing the Quality of Your Soil', we will discuss in more detail how much water different types of soil can hold. Sandy soils, for example, allow a lot of water to trickle into the deeper layers, meaning that the groundwater is well replenished, but the plants can dry out quickly despite high levels of precipitation. Loamy soils, on the other hand, retain moisture from which plants can benefit even over long periods with low rainfall.

Annual precipitation averages for the UK are 133 days of rain or snow, totalling 885 millimetres, with considerable regional differences. To find out the annual precipitation levels at your location, you can use the data from your nearest weather station. It should be valid for your garden with minor variations. It is well worth installing a rain gauge to record every shower. These plastic cylinders are calibrated so that each line represents one litre of precipitation per square metre.

Although you can now accurately monitor how much water your garden receives in terms of rainfall, this is still not enough to enable you to know whether your garden is sufficiently well watered. The last parameter to consider is the plants themselves.

Plants act like umbrellas and catch some of the precipitation in their leaves. The less it rains, the more raindrops remain, proportionally speaking, on the leaf surface without ever reaching the soil. It is only when it rains a lot that the excess runs off the leaves and moistens the earth. The strength of this umbrella effect, which in the technical jargon is called 'interception', depends on the type of plant.

Let's begin with the least favourable example: evergreen conifers, for instance the spruce. Their crown is so dense that even during a shower of 10 litres per square metre hardly any water reaches the ground. When the Sun comes out, this captured water evaporates again into the atmosphere unused. It's only if your rain gauge shows significantly more than 10 litres that you can rest assured that the ground also got a share. But the ground under the conifers can form a secondary barrier. There is often a carpet of dead needles beneath a spruce or pine, and this carpet, depending on its thickness, might also hold up to a third of the water that reaches it. The consequence is extremely dry soil, and it is not surprising that spruces and pines naturally occur only in the far north, where there is a lot of rain and where they do well.

Broad-leaved, deciduous trees allow much more water through their crowns, and in winter their bare boughs offer almost no resistance. There also tends not to be such a thick carpet forming on the ground beneath them, since leaves decay much faster than needles.

Let's look at the dwarves of the plant kingdom: grass and moss. While grass allows water into the ground even with a light shower of just a few litres per square metre, moss soaks the rain up like a sponge. Like spruces, moss returns the moisture to the atmosphere, forming a barrier that is only overcome by precipitation of over 10 litres per square metre.

The most favourable situation for water intake is open, unplanted ground – it is always immediately fully charged. However, this is undesirable in terms of protection against erosion. As a compromise, fortunately, there are also plants that promote water intake. Have you ever wondered why rhubarb has such large, funnel-shaped leaves? Watch what happens in the rain: the younger, upright stems catch the water and guide it down to their roots. The same can be observed in many other species (e.g. the dandelion).

In principle, foliage offers another advantage over bare soil: it slows down the force of the falling water and ensures a more even distribution. This is especially clear in deciduous forests: after all, the saying goes that it always rains twice in the woods. For hours after the actual shower, the water is shaken down from the leaves by the breeze. In this way, a heavy bout of rain is stretched over time, and the absorption capacity of the soil is not overwhelmed.

A rain shower can therefore have a completely different effect depending on the soil type and the plants.

Before we can tell once and for all whether your garden is getting enough water, a brief word about quantities: 10 litres per square metre sounds like a lot of rain, and indeed it has to be a good shower for this much to accumulate. It is the equivalent of a watering can full of water. In hot summers, that should be enough for your plants for a week, after which replenishment is necessary. But, if there is a coniferous tree looming over the flower bed, it will have been like a drop of water on a hot stone.

For this reason, you should drain the rain gauge after each shower. Adding together the quantities from multiple rain showers distorts the picture. An example makes this clearer. If there are two rain showers divided by a break of a few hours, both of 7 litres per square metre, all of this could get stuck in the crowns of conifers without a single drop reaching the ground. Meanwhile, in a heavy rainstorm where 14 litres of rain falls, some 4 litres might actually reach the soil.

You see, you can collect all the data you like about rainfall, but it is of little use if the vegetation in your garden blocks the rain from watering the soil. However, there is a very simple trick for checking whether your soil is sufficiently moist: push the humus to the side until you see the pure soil (the humus layer is usually just one to two centimetres deep). Hold a clump of soil between your index finger and thumb, and press them together. If the soil sticks together, it is moist enough. If, on the contrary, it crumbles as soon as you release your fingers, the soil is too dry.

You can carry out this stickiness test at various locations: on the lawn, in the beds or under trees. If you do it regularly after a rain shower and in dry periods, at different times of day and under different conditions, you will build up an overview of how much water your garden needs. Combined with checking the rain gauge, you will soon be able to estimate how much rain needs to fall before your soil soaks it up, and whether you need to help out with the watering can.

THE SLUG: THE GARDEN'S UNSUNG HERO

Whenever there's any decent amount of rain, they're off: slugs and snails are out on the rampage. Granted, like every gardener, I'm not a fan of snails in my vegetable patch. When my brand new courgette or cabbage seedlings disappear within days, or when my herbs and perennials – even the larger shrubs and bushes – are demolished the second they produce new leaves, my love for these invertebrates is sorely tested. And yet I don't see them as an enemy to contend with, I simply remove them from the beds and drop them in another corner of the garden. Like every living creature, slugs and snails also have their place in the ecosystem. They help spread fungi by dragging the spores around with them, and serve as dinner not only for the hedgehog but for many other species including firefly larvae, lizards and thrushes. If you see slugs purely as a nuisance and go after them with slug pellets or beer traps, their predators will also get it in the neck. And you risk accidentally killing a rare species like the red slug (*Arion rufus*). The red slug? Rare? It's true; this once very common species has been on the decline in recent years, but its fate attracts little attention due to the prevalence of a very similar-looking competitor. The misleadingly named Spanish slug (*Arion lusitanicus*), which in fact hails from France, has now spread all over Europe, displacing native species. It reproduces rapidly and occurs in dense populations with several slugs coexisting per square metre. Their bitter-tasting mucus doesn't appeal to the hedgehog and other garden predators, so it remains largely unchallenged. Unfortunately, the Spanish slug exists in various colours ranging from brown to orange and so can easily be confused with the red slug. The latter has become so rare that it is already on the list of endangered species in some regions of Germany.

Using beer traps or slug pellets means killing indiscriminately. Who knows? Perhaps your garden is a rare refuge for our vulnerable native herbivore.

A visual test is enough to determine how deeply the moisture permeates into the soil. Dark, damp earth stands out well against the underlying, drier soil. Mole hills are also useful for this: if I push a little bit of soil aside with my foot, I quickly see whether or not the ground beneath is dry.

How to water properly

If when you press a clump of soil between your forefinger and thumb, the stickiness test reveals dry, crumbly soil, then it's time to water your beds. Although the specimen only shows the condition of the upper centimetres, it is likely to be a similar picture deeper down, around the plants' roots. The underlying soil layers are not important for plant irrigation.

If it's too dry, you can of course provide remedial action by watering. It is worth first determining how much water your plants need by nudging aside some soil with your foot and checking how dry it is under the first few centimetres of topsoil. If it is just as dry, you should water a bit more.

The one thing you definitely should not do is water your plants every day. This is a sure-fire way to mollycoddle your plants. If overwatered, they grow used to always having enough moisture around their roots. In order to receive regular doses of water, they spread their roots out flat and close to the surface. Then woe betide them if there's a delay or you skip a day! They'll go on strike over the lack of water, drooping after only a few days.

Plants that have to make an effort to find the water they need tend to grow much deeper roots. They can always meet at least part of their fluid requirements, even when the topsoil has dried out. If you want to give these hardy specimens a helping hand in dry periods, then it is better to drench them thoroughly, giving them about 20 litres or two watering cans per square metre. This is easier with a hose, although this makes it trickier to measure the amount. A good way to work out an estimate is to fill your watering can using the nozzle or attachment that you normally use to water the beds, and measure the time it takes to fill it. Once you know how long it takes for 10 litres to come through the hose, you can calculate how long you need to water a particular bed. For example, imagine your bed is 30 square metres, and your hose fills a 10-litre can in 20 seconds. To achieve a dousing of 20 litres per square metre, you would have to water the bed for a total of 1200 seconds or 20 minutes.

This is no great chore if it means it saves you from watering the garden every day. In this way, you simulate natural precipitation and ensure the water reaches the plants' lower roots. The plant then has no incentive to direct its roots upwards. This drenching approach encourages resilience in your plants, and also allows you to occasionally go away for a fortnight without your entire garden dying of thirst.

Changes in animal and plant life

Rain is experienced in the same way by most animals as it is by us: it's cold, wet and therefore extremely unpleasant. If they have the option to, all creatures – whether they're insects, birds or mammals – will seek somewhere to shelter. Bees hurry back to their hive, blackbirds hide under tree canopies and deer seek shelter in the dense thicket. If there is no shelter, for example, for horses in open pasture, they turn their hinds to the wind, so that at least their faces are protected from the lashing rain.

Earthworms react quite differently. Traditionally known as rainworms, this name suggests that they are fond of soggy weather. But the opposite is the case: for them, every shower is a close shave with death. Their cosy tunnels, which are lined with mucus and stretch down as much as three metres in depth, fill up with water when it rains, meaning the worms face suffocation. To avoid this ghastly fate, they slither up to the surface as fast as they can. Another theory is that the thud of the rain drumming on the ground sounds to them much like a mole digging – their mortal enemy and feared predator.

Personally, I am more inclined to support the first theory, because it is certainly the case that after a heavy rain shower there are always several worms lying dead in puddles, suggesting that they cannot survive in water with its low oxygen content.

Deep within their tunnels, the earthworm cannot, however, tell what is causing that drumming noise. This is why they sometimes pop up between your wellies when you are digging over a patch of soil. Better safe than sorry, after all.

The arch nemesis of all gardeners also emerges in the rain: slugs and snails. These slimy animals would quickly dry out in the sunshine, which is why they spend the fine weather spells hiding in damp, dark places such as the compost heap. As soon as it gets wet, from spring to autumn, these slippery pests are off on the prowl. And so too are their natural enemies, the salamanders, newts, toads and frogs. When it rains in the evening and I'm out walking our old cocker spaniel, Barry, I have to tread carefully to avoid accidentally squashing a small amphibian.

Incidentally, a downpour often offers the best conditions for observing larger animals, too. The more they move about, the easier it is to spot them, after all. When it starts to pour, they (like us) hurry to find cover, under trees and bushes if possible. When the rain stops and the Sun comes out again, the soggy creatures also come back out into the open to dry off and warm up. Rural residents have a better than normal chance of spotting deer in the first few minutes after a storm. For this reason, if a storm is brewing, it's perhaps best not to cancel your walk in the country, but just to postpone it. You may be rewarded with a particularly impressive natural experience.

Plants also react to the rain with various physical changes, which can be suitable for predicting the weather (see page 7). Certain reactions are less helpful for forecasting, but are interesting nonetheless.

A classic example is spruce or pine cones. Children are often taught while out exploring nature that pine cones are like little weather stations. They open up in sunny, dry weather, making for a big, puffy cone, while in rainy weather they close up making the cone smaller and narrower. This much is true. However, since the change lags behind the weather, and this observation can only be made after the change in moisture level has already taken place, they are completely useless as a forecasting instrument.

An important change brought by the rain is that the plants get a thorough hose down. Leaves work like solar panels, so in order to function efficiently, the surface needs to be spick and span. The air contains a lot of dust, which accumulates on the leaves over the days and weeks, slowing down the plants' growth. It's only when it rains – and the heavier the better – that the solar panels are rinsed clean, and then they can shift back into top gear.

A common consequence of a rainstorm for plants is snapped stems. Whether out in the countryside where whole meadows and crops can be crushed by the rain, or in a domestic garden, where lovingly nurtured summer blooms are knocked to the ground, this phenomenon usually has a simple cause: fertiliser. Most plants, if left to grow naturally, are stable enough to withstand heavy rainfall without damage. However, our crops are bred to grow faster, and they are accelerated further by being doped with fertiliser or compost. This leads to tall, fragile stems that are insufficiently woody and therefore unstable. It is little wonder that these shaky structures fail under major stress.

During a rain shower, many flowering plants close their heads to protect pollen and nectar from being flushed out. After all, they need to lure the insects back as soon as possible once the Sun comes out, to make haste with the pollination process. If their goods are washed away, that's their chance of reproducing gone. This is a particularly harsh fate for annual species.

After the plants have finished flowering and the seeds have started to form, it's quite a different matter for many plants. Several species yearn for the rain at this stage. It helps to wash the seeds away and transport their progeny to new habitats. This is an example of ombrochory, where seeds are dispersed by rain falling on the mother plant. This makes a lot of sense in the case of the marsh marigold, because, as the name implies, it likes to grow by the water. But ivy-leaved speedwell, a common weed in many gardens, also makes use of heavy showers to send its seeds off on their way.

Sometimes, however, raindrops are merely the trigger for the plant to eject the seeds. Selfheal (*Prunella vulgaris*), also a common garden resident, disperses its seeds in seconds with the help of rain. The instant a raindrop lands on a leaf, it drips inside and pushes out the seed.

Reading hailstones

Besides lightning, the most fearsome feature of stormy weather is hail. I will never forget one occasion in our region, when a severe hail storm in July knocked about 70 per cent of the foliage from all the trees. Our vegetable garden was battered, and the lanes were covered with fallen branches and leaves. We made countless trips to the compost heap with the wheelbarrow, and the autumn harvest was pathetic. And yet I can't help having a certain fascination with this natural phenomenon, despite the adverse consequences.

Hailstones are formed in cumulonimbus storm clouds, where water accumulates around small particles and then freezes. In normal conditions, these lumps would quickly become too heavy and then fall to the ground as small snow pellets, or graupel. Within these tall, towering storm clouds, however, there are strong wind currents that pull these kernels up several kilometres, letting more and more water accumulate and freeze. Higher up in the cloud, the wind drops, so the hailstones fall down again into the stormy zone, where the updraught again thrusts them upwards. The more violent the thunderstorm, the stronger the winds and the more frequently this up-and-down process is repeated, and the longer it takes for the hailstone to become too heavy and finally drop to the ground. Small hailstones melt on the way down and splash onto the grass as particularly fat raindrops, while larger ones (up to the size of a football!) remain as lumps of ice on impact. Luckily, most hail falls within certain parameters, so that as a rule the largest hailstones are the size of a pea or a cherry at the most.

Each ice lump carries its (short) evolutionary story, written inside it. Large hailstones spent a relatively long time tumbling about in the thundercloud, while small ones had only a brief rollercoaster ride. If you cut open a hailstone and examine it up close, you'll see a layered structure, not unlike the annual rings of trees. And as in a tree trunk, the rings reveal a hailstone's life story: each layer comes from a single 'up-and-down' journey through the storm cloud, with five rings corresponding to five loops of the rollercoaster. We can also trace the relative duration of the roller coaster: short flights upwards leave a thin layer of ice, long ones a thick layer. Even 'long' flights represent a fleeting moment, in contrast to the rings of a tree which, in our climate, represent a year of growth.

In particularly violent upward winds, which can really keep the ice lumps hovering at the top of the cloud, just one up-and-down motion can result in large, heavy hailstones with no rings to be seen inside.

Evidence shows that hail can be hazardous when the stones exceed a centimetre in diameter, as the speed at which they fall also increases with the size. Smaller hailstones can rip through leaves, but plants tend to recover relatively quickly from such damage. From two centimetres in diameter upwards, hail can become a hazard for property such as cars, because the stones have enough mass to dent metal and smash glass.

One final tip: if you're faced with a particularly heavy hailstorm, naturally your first concern will be to check how the garden has fared. But before you do anything, pop a few hailstones in the freezer to look at later, after you've inspected the house and garden for damage.

Snow and frost

A snowflake is a little miracle – and each one is unique. The variety of possible combinations of the forms is so great that it's likely there have never been two identical flakes since the earth has existed.

In cold northern climates, snow will have a great influence on your garden's water resources. Winter is the season for refilling the tank: having dried out over the summer, the parched soil can now fill up again with rainwater before the vegetation starts pumping it back out again. Since plants hibernate in winter, or completely die back in the case of annual species, precipitation can soak into the soil unhindered (except beneath evergreen conifers), spreading down to the deepest layers. As long as the temperature remains above freezing, this drenching process continues without obstruction. Days on end of rain and snow – with sleet, puddles and muddy flower beds – offer the best guarantee for a productive growing season, so you should rejoice at every cloudy, drizzly day you are blessed with. We simply cannot get enough precipitation in many parts of Europe, since groundwater stocks have never completely recovered since the extremely dry heatwave Europe faced in summer 2003.

This refilling phase is sometimes disturbed by frost. If the frost lasts for several days, the soil freezes to a depth of several centimetres. When it rains as it gets warmer, the water meets an impenetrable layer of frozen ground and therefore runs off unused into nearby streams and rivers.

With snow cover, it is a very different story. The white carpet has an insulating effect that protects against ground frost. Fallen snow contains a lot of air, which blankets the earth very effectively against the cold. The thicker the snow cover, the greater the warming effect. The soil remains frost-free even at air temperatures of below 10 °C. If the temperature then rises above zero, the rain, together with the melted snow, can sink directly into the ground.

In the event of black frost, when there is no snow and low humidity, the vegetation is exposed to the low temperatures without this protective blanket. The ground is quicker to freeze and plants dry out. The only way to keep sensitive plants safe from the cold is to cover them with a protective layer of spruce branches or fleece, and to water them.

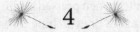

4

SUN, MOON AND STARS

NIGHT after night, you can observe one of the greatest wonders of nature from your garden, as long as the weather plays along. I have long been interested in astronomy, the study of the stars and other celestial bodies. In contrast to astrology, which deals with interpreting the future on the basis of the constellations, astronomy focuses on a scientific analysis of the universe.

A glimpse into the endless, dark expanse of space makes it clear how small and fragile our little rock called Earth really is. For this reason alone, it is worthwhile considering the celestial sphere of the night sky.

I find it especially fascinating to think that when we observe the night sky we are looking into the past. For the stars are nothing but very, very distant suns, whose light has taken centuries, if not millennia, to reach us. In the meantime, they have long since moved on, flared up more brightly or even burnt out altogether. The constellations, including the twelve signs of the zodiac used by astrologers to predict the future, may well have already regrouped themselves into quite different arrangements – heaven only knows.

If you have a camera that allows you to adjust the shutter speed, then you can capture a long-exposure image showing the rotation of the Earth. Mount the camera on a tripod and point the lens at a section of the night sky, ideally due north. Set the exposure for as long as possible, hours if you can. Sometimes there is the option for a remote shutter release, meaning the shutter is only activated when pressed again. The resulting photo shows the stars as curved lines arching through the sky, because the Earth (and with it your camera) has spun on its axis beneath them while the image was being taken. The longer the exposure time, the longer the tracks will run in an arc across the sky.

Cold nights and starry skies

When the air is clear, you can see up to 3,000 stars with the naked eye on a dark night. Which begs the question: isn't it dark every night? Well, no; in cities and residential areas where there are streetlamps, it clearly isn't: the artificial light blocks out most of the view. But the full Moon can also make it more difficult to see the stars, at least in that part of the sky.

When can we say it is night, after all? When the Sun goes down, it's dusk. The resulting half-light, or twilight, is the result of the Earth's atmosphere. The Sun is already beyond the horizon, but its light beams are refracted through the atmosphere and diffused into parts of the sky that are not directly in sunlight. This indirect illumination fades gradually, and depending on the season it can be one or two hours until it is finally pitch dark. It's only then that the faintest stars become visible and that the Milky Way unfurls its pale band across the sky. In June, when the nights are their shortest, this true darkness lasts only four hours.

And what of the stars? They are distant suns so far away that even in the most powerful telescopes they still appear as mere dots. But when your eyes get used to the dark (and this can take up to 30 minutes), the dots reveal something else: you might be surprised to find that many of them are coloured. Stars can be red, blue, yellow or white, depending on their temperature and the type of radiation they emit.

You'll see a lot more stars after a rain front has passed, when it suddenly clears up and the air is hosed down and dust free. If the sky is clear in the evening after a shower, this can be taken as a sign that the temperature will fall sharply the following night. Without the warming cloud cover, the evaporating moisture of the recent rainfall quickly cools the air close to the ground.

During a period of fine weather, on the other hand, a light haze often forms, which swallows the light of the fainter stars, especially near the horizon. In this case, poorer star-gazing conditions and a slightly hazy sky indicate good weather the following day.

The Milky Way is a remarkable sight. We are all part of an enormous disc-shaped galaxy containing hundreds of billions of stars, with many spiralling arms swirling around the centre. From our perspective, the Milky Way looks like a band streaking across the night sky, because we are looking through the disc from the position of our solar system in one of the spiral arms.

The Milky Way takes its name from its whiteish appearance. It is made up of an almost endless number of stars, whose light blends to a foggy blur because of the huge distance. Only the nearest stars, the 3,000 dots mentioned above, can be seen individually.

Together with the stars in our 'neighbourhood', our solar system (and with it you and me) is careering through space in a huge circular orbit around the centre of the Milky Way, at a speed of over half a million miles per hour. This corresponds to over half a million miles per hour. So when you gaze at the sky on a still night, consider the thought that everything is in fact racing around at top speed.

Shooting stars and cosmic rain

Shooting stars are the smallest heavenly bodies, if we can even call them that. They are regarded as a lucky omen, and to this day it's customary to make a wish for the future if you see one.

Shooting stars, or meteoroids, are bits of dust and rock that are hurtling through the universe and which burn up as they pass through the Earth's atmosphere. The glow as they burn up can be seen from very far away, depending on their size. You have particularly good chances of seeing one when the Earth crosses the orbit of a comet. Comets are just like huge dirty snowballs, consisting of ice mixed with dust particles. As the comet comes closer to the Sun, parts break off, forming its characteristic tail, which is nothing more than a trail of debris left behind as it crumbles. The most well-known intersection of the Earth's orbit with that of a comet occurs in early August, when we can enjoy the annual Perseid meteor shower. Hundreds of shooting stars are visible every hour: this is the debris shed by comet Swift-Tuttle.

Nowadays, it's easy to confuse shooting stars with wandering satellites. These are illuminated by the Sun and appear as tiny dots as they traverse the night sky, but at a second glance they can be distinguished from natural phenomena. Their trajectory is relatively slow, in contrast to how quickly a shooting star flares up and burns out. It is easy to follow the route of the satellite because you can track it until it disappears in the haze near the horizon.

Shooting stars cause a constant cloud of dust that rains down on us. This cosmic rain was, and still is, of great importance to the Earth (and, of course, to your garden): a considerable part of the water on our planet comes from larger orbiting bodies, such as comets, which smashed into the Earth way back in the mists of time. And even the tiny shooting stars add up to a staggering 10,000 tons that falls to Earth per day. A tiny speck of that is sure to land on your flower beds or vegetable patch.

Phases of the Moon

There are over a hundred different books on the subject of the Moon and gardening, so I will go into a few other aspects. It's undisputable that the Moon influences life on Earth. The evidence most commonly cited is the tides. The Moon and the Earth both orbit around a common centre of gravity. The Moon exerts a pulling force on the sea, creating a small bulge, or wave peak, of about 30 centimetres. On the side of the Earth facing away from the Moon, a second wave peak is formed, but here it is created by centrifugal forces – the outward pull felt by a spinning object like a carousel. As the Earth spins throughout the course of the day, the wave bulge moves across the surface of the Earth, always on the side facing the Moon, and so does the corresponding bulge on the opposite side of the planet. And so the water washes up higher on the beach or pulls away from the shore as this bulge passes by. The rise of the seabed towards the coast means that a minuscule wave bulge might be exacerbated by several metres, depending on the terrain, so that when the tide is in, for example on the North Sea coast, a stretch of beach several kilometres deep can disappear beneath the salt water.

Now, you might not think that your garden is affected by tides. Well, in fact, the Moon's gravitational pull tugs at not only the sea water but also the Earth's crust. Over the course of the day, your garden can bob anywhere between 60–80 cm up and down without you noticing it. These movements are on such a huge scale and so consistent that they can only be discerned with complicated measuring devices.

With such forces at work on the ground and in the sea, it is not surprising that many marine organisms use the Moon as a kind of calendar to spawn synchronously, the advantage being safety in numbers: predators cannot eat all the offspring if they all come at once. Depending on the species, it might be the full Moon, the crescent Moon or the new Moon (that is, complete darkness) that prompts the reproduction process in a given month of the year.

Might it perhaps be conceivable that even the organisms deep in the soil of our gardens use the daily rise and fall of the Earth as a kind of clock? Day and night, and the seasons of the year – which set the life rhythms for every species above the ground – have no impact on life a metre below the surface. Deep underground, it's always the same temperature and it's always pitch black. How do the thousands of species that live in these deeper layers order their lives? It may well be that the tidal forces are the only marker of the passage of time for these tiny creatures, but this has not yet been the subject of much research. And while many species remain undiscovered, let alone the focus of any scientific study, this topic may long stay shrouded in darkness.

There is still disagreement about the effects of the Moon on us human beings, but if even the soil of our gardens rises and falls every day, it is hard to imagine that our organism would be completely unaffected. After all, we have run our lives by the phases of the Moon since ancient times – hence the etymological link between the words 'month' and 'moon'. Even dates such as when Easter falls are pegged to the lunar cycle (the first full Moon on or after the vernal equinox). Whether menstruation is affected by the Moon remains unconfirmed (the menstrual cycle is usually between 28 and 35 days, while a lunar cycle lasts 29.5 days). A link was suspected in ancient times, hence the word 'menstruate' derives from the Latin *mensis*, meaning 'month'.

The planets
There are countless possible mnemonics to remind us of the sequence of the planets in our solar system, from 'My Very Easy

Method Just Shows Us Nine Planets' to 'My Very Educated Mother Just Served Us Nine Pizzas'. Each word's initial letter stands for a planet: Mercury, Venus, Earth, Mars, Jupiter, Saturn, Uranus, Neptune and Pluto. These planets all orbit around our solar system, with Mercury being the closest to the sun and Pluto the furthest away. The dwarf Pluto has, however, recently lost its planetary status in the eyes of many scientists: the tiny rock is simply too small to count. Once several other heavenly bodies of a similar size had been found and all of them categorised as asteroids (dwarf planets) or trans-Neptunian objects (TNOs), Pluto had to accept its demotion to this lesser status.

These asteroids, however, are not completely harmless. In 2011, the 400 metre-long object referred to as YU55 came closer to the Earth than the Moon – that is, in cosmic terms, a very near miss. If it had struck the Earth, huge swathes of land would have been devastated.

We cannot see such tiddlers with the naked eye, but we can see Mercury, Venus, Mars, Jupiter and Saturn without the need for a telescope. In contrast to stars, which emit light like our Sun, the planets are balls of gas or rock, illuminated by the light from a star, in our case the Sun. We can only see the planets of our own solar system in the sky, since the other stars are so far away that their orbiting planets merge optically into the same dot as the star itself.

It is relatively easy to tell the difference between planets and stars. Planets are particularly bright, as a rule, and their orbit always follows the same path through the sky as the trajectory of the Sun during the day. For this reason, in the northern hemisphere, you will never spot a planet in the northern part of the sky. However, if you are looking up at the sky in the southern hemisphere, such as in Australia, the opposite is the case: the Sun is in the north when it's at its highest point in the sky and it is the same with the planets.

Another distinction is the twinkling of the stars. Depending on the air turbulence, the stars often flicker and twinkle, but the planets appear as a steady source of light. This is because the stars appear as only the tiniest dots of light due to their extreme distance, while planets can be seen distinctly through binoculars as discs. This much broader beam of light is not disturbed so easily by slight air turbulence.

Because of their great distance, the planets probably have no impact on life on Earth. Even added together, they have only about one hundredth of the gravitational force of the Moon – not enough for any measurable impact.

FLORAL AROMAS FOR LATE RISERS

The flowers of different plants open up for business at different times of day. This helps spread the workload for pollinating insects and gives individual species the best chance to get noticed. A large group of flying pollinators – moths – leaves its visiting hours until after closing time. They are still hungry at this late hour, on the prowl for a sip of nectar. This is the opportunity for certain flowers to stand out from the crowd and beat the competition by flowering after sunset. Many species have adopted this strategy, such as evening primrose, which originally hails from North America. It is only at dusk that it opens its flowers and begins to give off its scent. Moths are lured in by the sweet message and land on the pale yellow sepals.

The colour yellow is typical of 'nocturnal' plants, because it is still relatively visible in the dark. But not all wait until the twilight to burst into action. Common soapwort (*Saponaria officinalis*) stays open all day long, but only starts to emit its seductive aroma after dusk. The same is true of the perennial phlox, whose pale pink blossoms are also easy to spot even at night.

If you like to sit outside on a warm summer evening, you'd do well to plant a few specimens of these night bloomers in your favourite spot. You'll be well rewarded by the opportunity to observe some shy visitors, including many you may never have seen before.

5

SUNSHINE AND DAYTIME

BESIDES our own planet, the most important heavenly body for us is the Sun. Its warming rays take eight minutes to reach us. This is due to its distance from us of about 93 million miles – a very significant distance. If the Earth were nearer to this furnace, it would share the fate of Mercury and Venus: all water would evaporate, and life would be impossible.

Inside the ball of fire that is our Sun, 500 to 600 million tons of hydrogen are burnt per second, radiating light, heat and other electromagnetic waves. But do not fear: despite this high consumption, there are still enough supplies to last for several billion years.

A scale model is helpful to get a sense of just how enormous the Sun is. If the Earth were the size of a cherry, the Sun would have a diameter of one and a half metres and would be 150 metres away.

In the daytime, the Sun's surface can be observed just as well as that of the Moon, because both appear as similarly sized discs in the sky (optically speaking, that is). But, unfortunately, the extreme brightness prevents us from looking at it for any amount of time. It's a shame we can't look at it more easily, because visible processes are taking place on the surface and they're not just pretty to look at. The Sun is sprinkled with solar spots, like birth marks on the whiteish-yellow face of our home star. Before I tell you how you can safely look at this impressive sight, I would like to go into what the sunspots mean for us.

Sunspots appear dark because these are the points where the Sun's surface is slightly cooler. Nevertheless, they are a sign of an overall increase in activity within the Sun: the more spots, the more radiation the Sun emits, and the warmer we feel. It's a cyclical process with more and more sunspots appearing over a period of several years, only for them to gradually disappear at the end of the cycle. These cyclical periods seem to last about eleven years, but needless to say the Sun does not always adhere to this schedule. After the disappearance of the sunspots towards the end of one of the last cycles (in December 2007), for example, there was an unusually long interval before they started to reappear. The surface remained immaculately pure, and even after years hardly any spots appeared. Experts reckon that the Sun's radiation is generally weaker at the moment and that the next sunspot cycle will be less pronounced, so that on average we will observe many fewer sunspots.

Europe has already experienced the consequence of decreased solar activity in the harsh winters of recent years. Freezing cold, frozen rivers and lakes, and here in Germany our railway network, Deutsche Bahn, has seen one train after another succumb to frost: all this is due to the absence of sunspots. And in all probability these are not the last extreme winters we'll face. The long-term fluctuations in our celestial heating mask the impact of the greenhouse effect, appearing to bring about a pause in climate change.

But for now, back to the garden. There is an easy way to follow the exciting action unfolding on the Sun. All you need are some binoculars and a piece of paper. But first, a warning: please *never* look directly at the Sun with binoculars, because your eyes would be severely damaged within seconds. For this experiment the binoculars are to be used as a projector, allowing you to safely view an image of the Sun projected on to the paper. To do this, hold the binoculars in front of the sheet of paper and point them towards the Sun (as though the sheet of paper wanted to look at the Sun). If you jiggle them back and forth a bit, you'll find you can project the Sun on to the sheet. To start with it will just look like a bright spot; then you can adjust the focusing wheel to bring it into sharper focus on the sheet. Now you can observe the Sun on the sheet and take a good look at the spots. It's more comfortable to use a tripod than to hold the binoculars steady, but more fiddly to set up. And again, please make sure that no one else looks directly at the Sun through binoculars.

If you regularly check on the sunspots and closely follow the Sun's activity, you can get a good idea of how cold the next winter will be, based on the ratio that the fewer sunspots there are, the colder it will be.

The course of the day

It is the rotation of the Earth around its axis (once in 24 hours) that defines the cyclical period we call a day. This may sound banal, but it affects our outlook. We tend to talk about the Sun as if we were still in the deep Middle Ages. We describe sunrises and sunsets, and portray the trajectory of the Sun in the sky as being from east to west; an alien hearing us speak would think we still believed the Sun was orbiting around the Earth. Of course, I also use these expressions; they're an integral part of our language. Nevertheless, I would urge you to try a small experiment at sunrise. Look eastwards, and remind yourself that the Sun is fixed. It is not the Sun that is rising, but the ground beneath your feet that is slowly turning towards the east. Every time I do this, I find it a very curious feeling. And yet this is what is actually happening, in contrast to the conventional way of looking at it. In the evening at sunset the whole thing works the same way, but rather than the Sun descending below the horizon, it is the horizon that rises up relative to it, as the Earth continues to turn towards the east.

ANTS : YOUR GARDEN HELPERS

When spring arrives in the garden, it's not just the flowers and perennials that burst into life, but also the less appreciated creepy-crawlies such as ants. Ants can certainly be annoying, no question. Not only are some varieties prone to giving you a painful nip, but they also love to turn your garden into an aphid farm. But their positive attributes are often unjustly overlooked. As they scuttle about running their complex civilisations, these busy insects also help break up the soil, making it looser and improving root penetration. And there's another reason that ants are in fact indispensable for many garden plants.

Have you ever wondered why certain wild flowers seem to wander around your garden? They're in one spot for a few years only to suddenly appear and start proliferating in another corner. It's the ants who are helping them spread their seeds. To encourage their loyal assistance, the plants offer them a small reward. Every seed has a juicy elaiosome attached: a fatty, sugary little morsel given in payment for the courier service. The ants haul this mouth-watering freight home for dinner, then transport the unwanted seeds up to 70 metres away to dispose of them. Both parties are satisfied: the ants are replete, and the plants can spread into new territory.

Plants which make use of this free delivery service include wild strawberries, dog violets, wild garlic, dead-nettles and forget-me-nots.

Clock time and true local time

What does your watch have to do with nature? Well, nothing actually, and that's exactly why we should talk about it.

A clock is supposed to represent the position of the Sun. This is also the reason why the hour hand moves around the clock face from left to right, like the trajectory of the Sun from the east (when looking south at the Sun, the east is to the left) over to the west (on the right). This is purely an optical illusion, of course; in reality it is our planet that is rotating.

Since you carry a handy astronomical instrument around with you on your wrist, you may as well put your watch to use for other purposes. It can be used as a compass if you're disorientated: if you point the hour hand towards the Sun, then south is always between the hour hand and 12 o'clock. (During the summer time, when our clocks shift forwards an hour, 12 noon should be replaced by 1 p.m.)

At 12 noon, the Sun should be exactly in the south and thus be at its highest position in the sky. *Should*. But remember that clock time is a compromise, and compromises are always flawed. The Earth is a sphere, and when the Sun is exactly at its zenith in Berlin, for example, it takes another 26 minutes for the Sun to reach its highest point at Cologne, around 360 miles west. This definition of time is called true local time or Local Mean Time (LMT) and it is, of course, different in every place. But if we used this, a country could not function. Nobody could make an appointment or write a timetable.

The problem is addressed with the use of time zones such as Greenwich Mean Time (GMT) or Central European Time (CET). It is only at the German–Polish border that CET, for example, corresponds to the actual position of the Sun, and if you should wish to match your watch up with the Sun anywhere else in Germany, you would have to subtract anything between a minute and over half an hour. In summer, an additional hour has to be subtracted, because the clocks go forwards an hour in spring.

It is worth checking the time difference for where you live. To do this, you need to know your longitude, which you should be able to find out from a map of your area (the coordinates are marked at the edges). There are various websites where you can enter your coordinates and work out your true local time, and therefore the difference between your LMT and your time-zone time. If your location is 15 minutes behind, for example, then you will need to add on 15 minutes to calculate your local time. The Sun would therefore be at its highest point, due south, at 12.15 (and in summer time at 1.15 p.m.).

UV exposure is also linked closely to the position of the Sun, which you can now determine from the clock: the time when exposure to UV rays is at its height is shortly before or shortly after 12 noon (e.g. with the delay of 15 minutes, as per the example above, this would be 12.15). Meanwhile, at 9 a.m. radiation levels are as high as at 3 o'clock in the afternoon.

The air temperature, however, doesn't follow quite this pattern. The Sun takes a while to heat up the air, so that the day's maximum temperature is reached only two to three hours after the Sun is at its zenith (i.e. at or around 3 p.m.).

The bird clock

Just as the clock can serve us for purposes in nature besides telling the time, so too can nature itself tell us the time, at least in a rudimentary fashion. As well as noting the position of the Sun, we should also observe the birds – and listen to their morning song.

Why do birds sing, anyway? It's certainly not for our sake that they fill the air with their melodies, and neither is pure *joie de vivre* the reason for their warbling. In fact, bird song is not unlike a dog raising its hind leg to urinate against the post of the local road signs: both serve to stake out territory. And since bird song is fleeting in nature, it needs to be repeated over and over. The basic message of their melodies for rival males is: 'Watch it! This is my patch!' Directed at females, on the other hand, the male's song is his way of promoting himself as a strong and virile mate. That is why most species don't sing in concert.

Species that put particular effort into producing a long series of sound sequences are also particularly staunch defenders of their territory. Blackbirds and robins are notable examples. House sparrows or rooks, which also count as songbirds, have a simpler call, and are more easy-going and amicable about nesting in close proximity to others of their species.

A diverse garden will become a habitat for a variety of species. But if all these birds were to sing at the same time, each one's melody would be drowned out in the cacophony of voices. In order for each singer to be adequately appreciated by his rivals or his sweetheart, each species focuses on a specific time in the morning. Or rather, not a time, but a certain position of the Sun. These are relative to sunrise, a precisely definable event. Unfortunately, it changes constantly, as throughout spring, the sunrise takes place a little earlier each day, until the summer solstice on 21 June when it starts getting later again. So, bird song is perhaps not ideal as a genuine replacement for your watch, although each species tends to observe its relative time slot, day by day, with astonishing accuracy.

You can find a good selection of bird species and the times they sing online. According to one German website www.biologie-wissen.info under the keyword *Vogeluhr* ('bird clock'), the schedule would look like this: the skylark begins while it's still dark, a whole one and a half hours before sunrise. The little redstart is next to step onto the stage. The blackbirds perform exactly one hour before sunrise, with the chiffchaffs following half an hour later. As the Sun begins to show on the horizon, then all the birds join in the dawn chorus. From then on, if you want to determine the time, you'll need some other living things to fall back on: it's time to turn your attention to flowers.

The flower clock

Carl Linnaeus, a Swedish natural scientist of the eighteenth century, made an exciting discovery during his nature walks. He realised that the flowers of different species of plants opened their flowers at different times of the day, with impressive reliability. They were so reliable, in fact, that they could rival the accuracy of the church clocks at the time. What better, then, than to create a living clock composed of a variety of flowering plants? Linnaeus put the idea into practice, planting a very special flowerbed in the Uppsala Botanical Garden. He arranged the plants in the shape of a clock face, dividing the bed into twelve segments. In each section, the flowers opened at the appointed hour, enabling passers-by to tell the time. However, the clock didn't function quite as intended, since the plants finished flowering after a few weeks and had to be replaced constantly. Moreover, specimens from the mountains behaved differently in the city, because of the warmer climate. Nevertheless, the principle is fascinating, and even without planting a clock-shaped bed, you can still tell the time from your garden with the help of your perennials and herbs.

Pumpkins and courgettes kick things off first by opening their flowers at 5 o'clock in the morning. From 8 a.m., the marigolds spread out their petals, and the daisies follow at 9. When the Sun is at its zenith in the south, midday flowers (*Mesembryanthemum*, also known as ice plants) open their blossoms. Between 2 p.m. and 3 p.m., dandelions start to close up, and by 3 p.m. the gourds have finished for the day. At around 6 p.m., poppies also shut up shop.

But why do plants go to the trouble of opening their flowers at different times? The reason for this is to attract pollinating insects, which risk being overwhelmed by choice. At the rush hour, when many flowers are open for business, the bees can't possibly visit all of the plants on offer, so some flowers would have to go without a visit. It gives you a competitive edge if you tout your nectar and pollen later in the day, when your rivals are asleep. So it's a way of improving the chances of pollination. It also helps the bees to take advantage of the available stocks of nectar, to bring as much as they can back to the hive for the winter. The more nectar supplies they bring back, the more bees survive of the next generation, which in turn ensures better pollination chances next year.

Having said that flowers keep good time, in fact researchers from the University of Göttingen have discovered that even flowers' internal clocks can be slow. This also appears to be linked to their effort to attract bees. Once the flowers are pollinated, they close on time. But if they're still waiting for a visit, they extend their opening times, probably in the hope they might still attract a passing pollinator. If you notice distinct changes from the normal rhythm of your garden's flower clock, there may be a lack of beekeepers and wild bees in your locality. You could take remedial action by setting up an insect hotel or even installing an entire bee colony in your garden.

The sundial

Where there is light, there is also shade. And this is the principle on which a sundial acts. Placed at the midpoint of a large semi-circular dial is a stick, known as a *gnomon*. When properly aligned with the points of the compass, the shadow of this gnomon passes over the clock face throughout the course of the day. The shadow's progress around the dial corresponds with the Sun's position in the sky, so you can read the time from the shadow cast on or between the numbers on the dial.

If you attempt to keep to time by this method alone, though, you'll find you're late to most of your appointments. Of course, a sundial tells us the true local time, as described on page 49. And a sundial can't account for the shift to Summer Time when the clocks go forwards an hour. If you want to read the exact time, you will have to add or subtract a few minutes depending on your position relative to your local time zone, and then add another hour during the summer. It might sound complicated, but as you only need to work out the difference once, you can grow used to reading a sundial relatively quickly. You could even turn the dial slightly, if possible, to tweak it to fit your calculations, since then the time indicated by the sundial would correspond precisely with the official time.

There is, of course, one advantage to this traditional type of timepiece over the more common devices we use to keep track of the time today: you only need to worry about the time when the Sun is shining.

6

THE SEASONS

THE Earth is tilted on its axis. That means nothing in the grand scheme of things, as there is no top or bottom to the universe, but for our climate and our perception of time it makes a world of difference. What we're talking about is the rotation of the Earth around an imaginary axis that passes through the two poles, and the position of this axis relative to our orbit around the Sun. One complete circumnavigation of our home star is what we call a year. In the course of this voyage, the northern hemisphere spends several months tilting towards the Sun (summer), then several months leaning away from it (winter). When the northern hemisphere tilts towards it, the Sun appears higher in the sky and gives us more warmth; while when we tilt away, the Sun lies lower in the sky and is correspondingly cooler. These fluctuations over the course of the year are what we experience as seasons.

The position of the Moon varies in a similar manner, but with the inverse relationship: when the north pole tilts away from the Sun in winter (and therefore more towards the night sky), the Moon appears lower on the horizon during the day and much higher in the sky at night. The same applies to the stars. Together with the long winter nights, this improved visibility is the main reason that winter is high season for amateur astronomers.

Since during our winter the northern pole of the axis is pointing away from the Sun, the southern pole is automatically directed towards the Sun (the axis is a straight line, after all). While we are shivering with cold up here in the northern hemisphere, it's high summer down on the other side of the equator.

Although the Earth's orbit is not quite circular, the variation in distance between the Earth and the Sun actually has no effect on the seasons. In fact, strangely enough, we are actually closer to the Sun in the northern hemisphere during winter than in summer.

In purely astronomical terms, the seasons begin and end on 20 March (the start of spring), 21 June (summer), 22 September (autumn) and 21 December (winter). The vernal equinox on 20 March and the autumnal equinox on 22 September are the two dates in the year on which day and night are equal in length. That is, it is exactly 12 hours from sunrise to sunset. Between these two dates, on 21 June, is the summer solstice: the day when the Sun is at its peak in the sky. The reason why summer only truly begins after this peak is that the Sun simply needs several weeks to heat up the air. This means the temperature increase lags a little behind the Sun's height in the sky, and the air temperature only reaches its maximum later in the season when the days are already getting shorter again. This is also why winter officially starts on 21 December, the winter solstice, and the weather gets colder only when the days are already getting longer. You can see the extent to which the temperature lags behind by comparing days of the same length. On 31 August, for example, we have as much daylight as on 11 April, but 31 August is usually much warmer.

But before we focus on each season in turn (where it's the temperature that interests us most), I would like to say a few words about frost. After all, the question of when we have had the last frost in spring and when the first might come in autumn is of crucial importance for many sensitive fruit trees and pot plants.

Ground frost

When water freezes, conditions become dangerous for many plants. After all, a large proportion of our vegetables as well as many ornamental plants come from warmer climates and are not equipped for minus temperatures. But even native shrubs and trees can be badly damaged by frosts in the spring, when the flowers and leaves are just coming out. Unlike tree trunks, wooded stems and twigs, a plant's fresh green growth cannot tolerate a cold shock and this part dies off. The plant might produce more green growth, but there will be no harvest that year.

As a gardener, therefore, every spring you are faced with the same questions. When can you plant out the flowers you've kept indoors in pots over the winter? When can the oleander bushes come out of the conservatory? When is it time to plant your tomatoes or courgettes out in the vegetable patch? Similar questions arise in autumn. The first frosts spell the end of the growing period, and with the overground vegetables, anything that is not harvested will turn into unappetising mush. If the temperature has been dropping, and if it is clear and dry during the day, then a frost will come as no surprise and you can prepare your garden accordingly. You have time to pick the last crops, bring the potted plants into the conservatory or the house, and cover delicate perennials with garden fleece.

The real danger is with an unexpected frost. This can happen when there's a combination of damp soil and a sudden change in the weather. A very typical situation is when a rainy warm front passes and clear skies follow. The clearer weather is the cold front of a low-pressure area, and as the clouds disappear, the atmosphere loses its warming blanket. This allows the temperature to drop sharply at night. The moist soil also contributes to a cooling effect, as the water from the previous rain shower slowly begins to evaporate. Sometimes this leads to there being so much water vapour in the air that clouds of mist rise up from meadows and forests. And this process consumes energy which is drawn from the air above the ground. This causes the ground to cool by several degrees, which is often crucial in spring or autumn. Although your thermometer might show 4°C, the flowers in the beds are in fact freezing. The problem is that thermometers are generally placed at eye level. Instead, to get a true reading of the temperature on the ground, that's where the thermometer needs to be placed, or else we need to calibrate the reading to allow for the elevation of the measuring device. It's worth remembering that the values we see in the daily weather reports are based on measurements taken at a height of two metres above ground level at the weather stations. Strictly speaking, there should be a specific weather report for our gardens, but because the weather forecasters know our dilemma, they do at least usually issue warnings of ground frost.

Whether your garden is at risk of frost when the air temperature is 4°C depends on the layout: if it is open to the wind, with exposed lawns and beds, you should take care to protect plants that are sensitive to frost. If, on the other hand, it is surrounded by established trees and hedges, it may be that it can be two degrees colder before you need to take action. The shelter and wind protection mean that the ground is considerably slower to lose heat. This effect can also be observed in cars parked under trees: the windows don't freeze as quickly as on cars parked in the open.

It is not always clear whether a plant is frost resistant and can remain in the garden, as not every specimen of a species behaves the way it says in a textbook. For you as a gardener, it is of course extremely annoying when a supposedly frost-resistant plant suffers frost damage. If this happens, you shouldn't necessarily complain to the garden centre about misinformation, because there are a number of reasons why hardy specimens can come to a chilly end.

The first possibility concerns exotic plants. Many plants we have imported from far-off lands inhabit regions with a similar climate to ours in Europe, and these can safely be left in the garden all year round. The climatic conditions are usually roughly the same as those of their original homeland, but it is still only approximate. It may be that your imported plants start their winter preparations two weeks later than native species. In normal years this is unproblematic, but if winter comes early one year, it can catch these exotic specimens on the hop.

The second possibility concerns native species. They should generally have no problem coping with severe winters (otherwise they would already be extinct). Nevertheless, they are sometimes caught out. Minus temperatures can cause grief for young trees and shrubs, which are used to growing up under the protection of their parent plants. Besides, in more natural settings, the soil is covered with a thick layer of humus, which keeps the earth warm.

Many gardens, on the other hand, tend to have a climate more like open ground. In clear, long winter nights, the wind dies down and the ground radiates all of its warmth up into the atmosphere, unhindered. As the layer of air directly above the ground emits a lot of heat, in particular, it also has a disproportionately quick cooling effect. For your garden plants, this means that the smaller they are (and therefore lower to the ground), the more they are at risk. As young saplings, they have to endure temperatures up to ten degrees lower down there at ground level than their boughs do when in a taller, mature state. Of course, even though shrubs and trees increase in size, the base of their trunks or stems remain on the ground. So why do mature specimens not freeze? It's because of their wooden stems. Wood has a fibreglass-like construction, making it very tough and resilient, yet with a certain flexibility. Even when exposed to freezing temperatures, it doesn't crack and no cells are ruptured. In the case of thin branches, this process of lignification (turning woody) is not yet complete: they are not yet toughened up by a woody outer layer and their softer tissue can easily become damaged by a frost.

Young specimens are protected from the frost if they grow up under the shelter of older trees and shrubs, where their protective canopy prevents severe cooling on clear nights. Out in the open the conditions are much harsher. Nevertheless, even in harsh winter conditions your young trees and shrubs have a good chance of survival if you give them a helping hand.

To prevent the ground from losing heat, it is usually enough to lay a thin fleece cover over the plant – as you would for protecting perennials. This increases the temperature of the surface by several degrees – an increment that can mean the difference between life and death.

The most frost-sensitive parts are the tender, young shoots. They are still growing, and the cells within their branches are soft and green. It takes several weeks for the lignin to be formed and deposited in the cells, to harden the stem so it can withstand freezing. In order to get everything done and dusted long before the winter, trees and shrubs stop growing taller in the summer and start working internally towards improving stability. Applying too much fertiliser (nitrogen fertilisers are especially dangerous) can wreak havoc. Doped up, the plants just focus on growth at the expense of stability, so that in autumn they are not ready to start lignifying their cells in time – that is, turning them to wood. A hard frost overnight is enough to freeze the immature shoots and turn them brown. This can spell the end for a young specimen. So be sparing with the fertiliser; better for your plants to grow slowly but surely.

But frost isn't all bad, and I should say a few words in its defence. Our native flora needs it, after all, or more precisely, it needs the forced rest that it brings. This is also the reason why trees from Central Europe cannot be kept indoors all year round as bonsai pot plants. Just as we need our regular sleep, so do plants of the temperate latitudes need to slumber for a few months. It is only in the spring that they have the strength to produce new shoots and to grow again.

In certain, very specific weather conditions, nature offers us a marvellous, highly unusual spectacle which goes by the name of 'hair ice'. It is formed on dead branches of many species of broad-leaved trees, lying on the ground, and looks like a fur of the most delicate ice threads. These 'hairs' are very close together and can be several centimetres long. If you touch them, the fluffy structure melts into pure water.

This rare phenomenon is caused by the fungi working away inside the branches. After a relatively warm, wet period in winter, the sky sometimes clears at night, and the temperatures fall below freezing point. Warm and wet – these are the ideal conditions for fungus. Inside the wood, the fungus is well insulated and it produces some warmth at the same time. Fungi breathe as they function, and this exhaled breath escapes through the pores in the wood, where the vapour immediately freezes. The hair ice becomes longer and longer until the wood cools down and the fungi stop functioning. In the morning, pieces of deadwood can be discovered, coated in this white icy fur, but their fragile splendour collapses with the first warm rays of sunshine.

Spring

When winter seems to be slowly fizzling out with its transitional weather of chilly drizzle, I can hardly wait until the gardening season begins again. At long last, I start to spy movement among the plants, there's a trace of changes afoot, and finally the first warm days in March pave the way for a coffee break outside on the patio.

When our children were small, they would ask me over and over, 'Is it spring yet?' As if the official chronological start of spring would necessarily involve a change of weather that would put an end to the bitterly cold days. There are various definitions for the beginning of spring. As we saw on page 56, astronomy looks at the earth's progression around the Sun and defines spring as starting on 20 or 21 March in the northern hemisphere. This is the date when day and night are exactly the same length, all over the globe. From now on, the Sun rises to a higher position in the sky over the course of the day, and warms the northern hemisphere with renewed strength.

Meteorologists, on the other hand, define 1 March as the beginning of the spring, because they divide the seasons into whole months, and for many species of plants March already lies within their vegetation period, their active phase. And the third definition is that of horticulture: it defines spring in terms of the life cycle of plants. If a certain plant is active, then for that individual species, spring has sprung. This way of looking at it of course means we have thousands of possible springtime springboards, reflecting the diversity of species in the region in question. It's not just latitude and altitude that affect the timing of the start of growth; even within one small region, over a distance of just a few miles, the start of the vegetation period can vary vastly depending on the local microclimate.

Despite this diversity, I believe that the final definition of spring is the most helpful for gardeners. After all, what use is it when the calendar says that it should be warmer now, but the garden is still covered in snow? In terms of preparing your beds, it is much more important to know how far along in the season your local microclimate is. And you can look at certain exemplary plants to follow this progress.

Experts divide the seasons up even further, in order to chart the progress of the vegetation period more closely. These subdivisions are based on an idea developed in 1955 by Dr Fritz Schnelle, a German agrarmeteorologist – that is, a researcher who uses the study of weather and climate to improve agricultural production. Linked to this is phenology: the study of plant and animal life cycles, and how these are influenced by seasonal variations in climate. With reference to the seasonal characteristics of certain plants, Schnelle developed a schema of ten 'phenological seasons' of the year: the active seasons of spring, summer and autumn are each divided into three periods, and the tenth is winter.

The pre-spring often begins while it's still officially winter according to the calendar, with the snowdrops that usher in the start of the garden's year. Together with the dusty tails of hazel catkins, the snowdrops remind us that we can now start getting the beds ready. The first vegetables, such as the broad beans, can even be sown outside.

We speak of early spring when the currant buds open. Sloes and cherries follow soon after with their blossom, but the leaves come later.

The apple blossom marks the coming of full spring, and now it is also warm enough to sit and drink a cup of coffee outside in the garden.

There are various websites in Germany where you can track the spread of apple blossom across the country from the south to the north, just as in Japan they track first sightings of cherry blossom (for example, on the website http://www.jnto.go.jp/sakura/eng/index.php).

When the apple blossom has finished, garden owners are impatiently waiting for the summer, but it's not yet plain sailing as there still remains the risk of one last winter greeting: the oft-recurring cold snap known in European folklore as the feast days of the Ice Saints. The Ice Saints (*Eisheilige* in German, *les Saints de Glace* in French) is a name given to a group of saints whose feast days fall on 11 to 15 May, and traditionally this week is noted as being prone to the last risk of frost. Many people wait until this week has passed before planting out plants that are susceptible to freezing. But how reliable is this traditional way of anticipating the last frost? Unfortunately, not at all, because in many years the particular weather conditions that lead to a late frost never come about, whereas in some years it occurs unexpectedly at the end of May. Besides meteorological activity, altitude and topography also play a role in the risk of frost. It is, of course, often colder in the hills than in the valleys, meaning that the Ice Saints will strike here much later in the season. Where we live, at about 500 metres above sea level, we regularly get frosts again in early June, so we have had many flowers succumb to the Ice Saints' icy spell (we are as impatient as anyone else when it comes to planting out spring flowering beds). Certain inauspicious valleys, which see the cold air flow through from the surrounding hills, can also experience similar 'unseasonal' cold snaps. It is only the low-lying plains and large river valleys, where the water contributes to heating the ambient air, which tend to be frost-free after the Ice Saints' week in early May. Nor can we rely on the effects of climate change: late frosts are becoming rarer, but they still remain a possibility.

As spring comes to an end, for animals and plants much of the hard graft has already been done. Trees and shrubs have produced leaves and long, green shoots. Grasses and herbs have shot up and flowered; the first plants are already bearing fruit. Meanwhile, in the nurseries of the animal kingdom, this year's offspring have already grown up and left home. In the coming months there is still work to do: shoots to make woody, seeds to grow from flowers, and some animals may rear a second or even third generation of little ones. However, all of this is child's play compared to the early spring, when animals have to defend their territories before they can reproduce. For plants, too, forming fruit is a much simpler task than the painstaking start after the winter, when they have to use their very last reserves of energy to burst back into life. Nor are plants spared the need to defend their territories: a spot in the Sun is a rare thing to be fought for. When spring comes to an end, this furious time of exertion is over, and the warm season can be used to complete, in relative peace, what was begun in the spring. Perhaps we humans do this too instinctively, as life settles down to a gentler pace; it's surely not only because of the Sun's warm rays that a summer garden is so relaxing.

Summer

From a purely scientific point of view, the hot season starts on 21 June, the longest day of the year (the astronomical definition), or on 1 June (the meteorological definition). But let's stick with our plants' perspective. The start of summer is marked by flowering grasses in the meadows. A little later, the tractors are out harvesting the hay, while in our hedgerows the elderflower is blossoming. If you have strawberries you might be picking them for weeks now, and your vegetable plants will be growing at an astonishing rate. The potato flowers mark the transition to high summer. Most of the crops in the allotment have now grown small fruits, which flourish splendidly in the warm Sun.

For the animal world, summer is the time of great bounty. There is food in abundance, and the predominantly dry, warm weather supports the growth of healthy offspring. There is relative peace in which the young can be taught the serious business of survival – and yet the cold season comes round fast enough; eight out of ten young animals will not live past their first year of life.

When the rowanberries turn red, summer is coming to an end. But even when autumn seems to be here, we sometimes have a happy return of summery conditions: what we refer to in English as an Indian summer. This is the time of year when you can see thousands of young spiders bungee-jumping off into the distance on their silvery threads. The sight of their wispy threads, like an old woman's grey hair, gives the German name for an Indian summer: the *Altweibersommer* ('old woman's summer').

Autumn

When the leaves change colour, it's harvest time. It's only now that the seeds and fruits of most plants are ripe, since they take as long as they can to store away reserves in the form of oil, sugar or starch. This is especially important for annual plants, which can only survive the winter in the form of seeds. All other species store their winter reserves in their shoots or roots.

These efforts can also be seen in your vegetable patch. Carrots, potatoes or parsnips all store starch and sugar in their 'cellars'. The goal is to be able to burst suddenly into life again in the spring, an energetic process that will eat away at the reserves for weeks. These biennial plants have a clear advantage over annual plants, which must first germinate from seeds every year and whose seedlings have to very quickly produce their own energy for growth.

Grass also accumulates nutrients in its dense roots, and this is one reason why voles manage so well throughout the winter: they love the nutritious strands that thread through the upper layers of the soil.

We know autumn has come when meadow saffron, also known as naked ladies, is blooming (and indeed the other common name for *Colchicum autumnale* is autumn crocus, although it is not related to a true crocus). On clear nights there may be the first ground frosts; delicate potted plants should now be tucked into a sheltered spot next to the house or brought into the house or conservatory.

When autumn is in full swing, it's the harvesting season: time to pick apples and dig up potatoes. After that, the rest of the vegetable patch can be cleared, leaving only the frost-hardy species such as cabbage, chard, winter radish and parsnips.

Shrubs and deciduous trees pack away their solar panels: they shed their leaves because they hibernate through the winter and take a break from photosynthesising. Deciduous trees and bushes discard their now superfluous foliage, to provide the autumn storms with as little surface as possible for their onslaught; with full foliage, a tree is not unlike a sailing ship which catches the wind in its sails. This last phase, late autumn, paves the way for winter.

When the temperatures drop, a whole host of bird species begin their journey to the south. They flee not only the frost, but also the food shortages associated with the snow. True migratory species all migrate to warmer climes, with just a few exceptions, whereas 'partial migrants' may choose to stay put for the winter. Sedentary birds, or resident birds, are those non-migratory species which remain with us all year round.

Bird migration has long been an enigma, and is generally understood to be instinctive, genetically programmed behaviour. But it seems that cranes and other migrants are not hard-wired to do it; they could just as well decide not to. If they do go, they decide when they are ready to commence the big journey. And this decision-making freedom is another natural phenomenon you can use for your own weather prediction.

Cranes and wild geese, as well as many other species, don't make a pre-emptive escape for the south, and neither do they stick stubbornly to the calendar. The only aspect that is genetic is the motivation to leave when they get uncomfortable, their *wanderlust* instinct. What makes them start their journey is a change in the weather. If it is suddenly uncomfortably cold, if it starts snowing heavily, for example, they remember that they don't have to stay put. 'That's it,' they think. 'Let's get out of here!' If, on the other hand, it remains warm and rainy, if they can still find enough food in the fields and meadows of the north, then they'll postpone departure. Delaying their departure in the event of warm weather also has another quite practical reason: mild temperatures mean south winds, which transport the warmth from southern Europe to the north. The south wind, however, would be a headwind for migrating birds, and to fly in such conditions would be exhausting. Cold bursts, on the other hand, are usually accompanied by brisk winds from the north, the tail wind making the perfect vehicle to carry migrating birds to the south with relatively little effort. Large flocks of cranes, or other migratory birds, suggest cold bursts in the north, and as a rule winter is not far off.

In the spring, it's the other way round: high temperatures in the south prompt the birds' departure, as the flight to the northern breeding grounds is easy with the warm southerly winds behind them. The arrival of the migratory birds is therefore a relatively reliable announcement of the start of spring. However, I wouldn't rely entirely on the migration of cranes as a predictor of the weather, because sometimes a flock can be caught out by a change in conditions and forced to make an unplanned stopover.

It is especially exciting when a rare guest, such as a twite, makes an unexpected visit to your garden. In Britain, these brown birds, the size of a great tit, breed in moorlands, including the Scottish Highlands and the Pennines, and winter in the lower lying coastal fields and saltmarshes of eastern England. In Europe, they come south from the taiga (the boreal forest) for the winter and don't usually migrate much further south than the coasts of the North Sea and the Baltic. Heavy snowfall, however, can prompt them to make a detour into the interior of mainland Europe, so they can sometimes be seen landing in domestic gardens further south in Germany, for example. This can be another indicator that a harsh winter is on its way. The same can be understood if you spot a cinnamon-breasted bunting, a waxwing or a nutcracker in your garden.

In the Middle Ages, the Siberian jay would make an occasional foray from the northern taiga down to Central Europe. At that time, people were well aware of the omen these grey-brown birds brought with them: a very severe winter.

And as the harsh winter weather drastically exacerbated the plight of the population, this messenger from the north acquired its distinctive German name, *Unglückshäher* ('calamity jay'). This echoes the bird's Latin name *Perisoreus infaustus*, where *infaustus* means inauspicious.

As their starting point and destination is not genetically programmed in migratory birds, it is no wonder that climate change is already prompting tangible changes to migration patterns. In some years, the last cranes set off for the south just a few weeks before the first flocks are already heading north again. Some species, such as the black redstart and song thrush, have in many cases given up migrating and have started to spend the winter here in Germany.

Just as with spring, there is no way of making long-term predictions as to when the first frost is likely. If you pay close attention to the weather and prepare your garden as outlined in the earlier chapter on frost, you can at least protect your most susceptible plants from any nasty surprises overnight. Unfortunately, there is no rule of thumb in this respect: one year, the temperature may drop below freezing in September, while in other years it won't happen until November. Climate change will cause a delay in the first frosts, yet we cannot ever rule out the surprise of a particularly cold spell.

Winter

I like winter. Outside, the cold wind howls around the house and the landscape sinks beneath the snow, and inside it's cosy and warm by the fire. The only downside is that I hardly spend any time in the garden at this time of year.

For the plants it is a very different matter. Winter is the harshest time of year, when all growth ceases. Your garden is effectively a desert during this period. If the temperature dips below zero, this cuts off the water supply: as far as the plants are concerned, it's as arid as the Sahara. It is no longer possible for plants to react to anything, including to an infection or injury. The trigger for this cessation of activity is the frost. For trees and shrubs, herbs and perennials, it is as though the chill air freezes the blood in their veins. And as ice has a larger volume than the equivalent amount of water, the cells can rupture.

To avoid death by frost, plants have evolved various strategies. Many species have dispensed with complicated frost-protection techniques and with it the chance of a long life. Such plants – annuals – freeze to death with the first overnight frosts. They live through the cold season only in the form of seeds. As these contain hardly any water, they can survive minus temperatures unharmed. In order to be able to burst back into life in the spring, each seed contains reserves of energy in the form of oils and fats. And this is also the disadvantage of this method of overwintering: from November to March, hungry birds and mammals are hot on the trail of these kind of tasty treats and thus most of these tiny power packs are gobbled up before winter is over.

Hibernation as a mature plant, as practised by woody trees and bushes, is a slightly less hazardous approach. This has the advantage that the plant does not have to start again from square one every spring, but can grow a little more every year. In order not to freeze, most of the water needs to be pumped out of the trunk and branches. The plant's cells are also made robust and resilient by lignification, so they do not burst if they freeze.

Another strategy is pursued by those perennials that bid farewell to their over-ground part: like deciduous trees, they withdraw reserves from their leaves, but herbaceous perennials go a stage further and let the green stems die back, too. During the summer, nutrients are stored away in the roots, so that enough energy is available for them to grow back in the coming spring.

Perennials have a decisive advantage over annual plants: they can rocket into action as soon as the temperature warms up in spring and achieve a height advantage over the annuals, which have to start as tiny seedlings, gaining height slowly and with great effort. However, their underground storage is particularly at risk from predators during the winter: they are a sought-out delicacy for voles, who feast on this underground treasure trove.

Winter is the hardest season for animals, but not because of the cold. After all, they can wrap up warm, growing a thick winter coat or denser feathers. The rich foods available in autumn help them grow a thick, insulating layer of fat under the skin to brace themselves against the chill. And, if they need to, they can simply sleep through the worst months. Since plants pack up and stop producing supplies, there is hardly anything for animals to eat, after all. It's a little easier for the carnivores, at least to start with, for whom the offspring of many animal species, grown nice and plump during the summer time of plenty, now offer rich pickings. Their youthful inexperience makes them an easy prey. Indeed, on average 80 per cent of wild animals lose their lives in the first year.

So, the frost is not always a problem. All animals have coping mechanisms, including insects. Contrary to popular opinion, a harsh winter has little effect on their population size, otherwise most insect species would have long since become extinct. A particularly cold winter is therefore no guarantee that there will be fewer mosquitoes, ticks or other pests in spring.

What is much harder for animals to bear is the cold, wet weather. At temperatures slightly above freezing, rain or mist causes the body temperature to fall quickly. Even for us humans – who can easily wrap up with extra layers of clothing – damp and cold together make for the worst combination. Water conducts heat better than dry air, so the body cools faster. For animals, this means that they consume more energy to maintain the required minimum temperature. And if their fat reserves are depleted too early, before the end of the winter – they're done for.

What does winter have in store?

In some years, the trees seem to be overloaded with fruit. When I lead tour groups through our woodland reserve in autumn, I am often asked whether the unusually abundant supply of acorns and beechnuts suggests a severe winter is coming. I'm afraid my answer tends to disappoint: sadly, no, we can't predict the future from a bountiful crop. After all, the fruit buds are produced the previous summer. If it indicates anything, an abundance of acorns and beechnuts in the forests suggests a stressful summer the previous year; that is to say, a dry one. Plants react to stress by ratcheting up reproduction, because they fear that if they are exposed to any further strain, they will not be much longer for this world.

The squirrels and jays, who seize on this bounty and stow the nuts and seeds away in countless winter stashes, are clearly no more reliable a means of predicting the weather in the coming months.

Statistically speaking, a more certain indicator is the correlation that shows how long periods of good weather in autumn are often followed by a severe winter. The arrival of Nordic migrant birds, described on page 68, is also a relatively reliable indicator of rapidly dropping temperatures.

But, as we all know, winter is not just about cold and snow. At least as important are the frequency and strength of storms. In recent years, many hurricanes have raged across Central Europe, hitting our woodlands and forests hard. Not only have trees fallen in their millions, but the damage has been evident in toppled pylons, de-roofed houses and even human casualties.

The frequency of storms is linked to the occurrence of wintery low-pressure areas. Normally, the winter in mainland Europe is characterised by highs that bring clear skies and brisk, cold air. But now and then, these highs are pushed aside by a small low-pressure area, decking the landscape with snow.

A low-pressure storm front is more typical of the transitional periods, i.e. autumn and early spring. In recent years, however, there has been increased movement of weather areas, so that in the winter we also find ourselves in a strong low-pressure area. Such circumstances can mean one storm after another for weeks.

We can make a short-term prognosis about the likelihood of storms by looking at the prevailing weather conditions. If our usual winter high is established here in Central Europe, then the storms tend to bypass us. If, on the other hand, from early on we experience one low front after another, then we're in for a turbulent winter with driving rain and fierce winds.

7
LIVING WITH CLIMATE CHANGE

For every living being, climate change has its own significance. Some species are Sun lovers, after all. They can never be warm enough, and every sunny day is celebrated as a blessing. What is true of us humans is also the case in the animal and plant kingdom. Various species that prefer warmer weather are gradually migrating further to the north, as new habitats open up that were previously unsuitable due to low temperatures or excessive dampness. Besides insect species such as the Asian tiger mosquito, deciduous trees are also spreading further into the far north, into regions previously reserved for coniferous forests.

The main concern for our native plants is the question of water supplies. High temperatures are becoming more and more prevalent, even without the greenhouse effect. The warmer it is, the thirstier you are; the same is true of garden plants. But before we take a glimpse at what the future might have in store, I would like to reveal some of the strategies that nature has developed for the economical use of water.

Good water management

Plants can regulate their water consumption. If enough moisture is present, photosynthesis can run at full steam in the warm season. Trees are particularly large consumers of water. In contrast to bushes and perennials, they have a much greater surface area of foliage and thus also a much higher evaporation rate. On a hot summer day, a large, mature deciduous tree consumes up to 400 litres of water. With most trees, the roots stretch much deeper than other plants in order to access the moisture of the unused layers of soil. Despite this advantage, it's clear that the underground reserves can be very quickly used up. If there are not sufficient top-up supplies in the form of rainfall, the tree needs to learn to rein itself in, to avoid dying of thirst. The underside of the leaves are covered in tiny pores called stomata, which look like minuscule mouths. They function in a similar way, too: this is the organ through which the tree breathes. And just as we breathe out, and lose water as we exhale, so do trees (except that their exhaled breath contains a lot of oxygen). When they sense that it's getting too dry, these pores start to close up. Water consumption levels can be greatly reduced by this means, but the tree's ability to photosynthesise is now also considerably reduced. If this happens over a longer period of time, trees produce less growth and less fruit. Alarm bells should now be ringing if you're relying on the fruit harvest, because in such conditions, apples and pears, for example, will be stunted in growth or discarded from the tree altogether. If the need for water is still too high, a tree can also discard some of its foliage. This emergency measure is often observed in hot summers in July.

If your garden experiences regular periods of drought, the trees can learn how to consume water sparingly. They grow much more slowly, but perform significantly better in heat waves than trees that have been 'spoilt'. In the reserve that I manage, I have noticed that in dry seasons, trees that have been spoilt have died in normally well-watered locations, while specimens that are used to thirst cope well with a crisis period in exceptionally dry conditions.

Grass is not so good at economising water and can't cut back its water consumption anything like as easily as trees. This is why the blades of grass wither during prolonged drought, and the lawn starts to turn an unsightly yellow in patches. This is a good time to check the soil: wherever the lawn begins to yellow, the water storage capacity of the soil is particularly low. (See page 87 for more on soil.)

Grass does, however, have its own survival strategy. The plants live on in their dense roots and start again with new shoots after the next rainfall. This is also the reason why it is normal to see your lawn turn yellow during a hot spell; it will recuperate at the next rainy opportunity. It is therefore unnecessary to water, unless it's especially important to you that the lawn is always green.

Garden shrubs and perennials, on the other hand, are able to economise as well as trees; they are also capable of learning to be frugal and to go carefully in times of drought. This ability to learn has not yet been lost in them by breeding, because our cultivars (cultivated varieties) are still very close to the original, wild varieties. They need to have this capacity, after all, because they are supposed to survive for many years in the garden and, as a rule, without care being lavished on them. It's a different state of affairs with vegetable plants and annual summer flowers. We have already discussed, in the chapter on rain, their fragility, which can be traced back to breeding and over-fertilisation. These plants are bred for high yields or the quality of their flowers. The wild species is often only faintly recognisable in the cultivated varieties, and the price for their incredible yields is the loss of certain original characteristics. Thus, courgettes and cabbages are now so sensitive that the hot

PLANT GALLS

Strange lumps and bumps can sometimes appear on the leaves of shrubs and trees in the summer. These distinctive growths can be round pod-like structures or even phallic protrusions, and certain varieties are also hairy. It's hard to tell from their mysterious appearance, but these are the work of parasites who have unceremoniously hijacked the leaves. Substances they secrete force the host plant to produce little dwellings that provide the parasites with shelter, food and protection from predators. If you open one up, you will find a tiny grub inside. This is the offspring of a gall midge, gall louse or gall wasp, and the larval stage inside this cavity makes up the greatest proportion of its existence. The little larvae then pupate over autumn, when the leaves wither and fall, and emerge from the pupae in late winter or spring, depending on the species, to begin their short life as an adult. They no longer feed in their mature state, but focus singularly on laying eggs on their favourite plants. Then they die.

Typical host species include oaks (the oak gall wasp), beech (beech gall midge), spruce (spruce gall louse) and roses (the wasp *Diplolepis rosae* whose gall is variously called robin's pincushion, bedeguar gall or moss gall).

These creatures rarely cause the plant serious damage, with trees and shrubs usually able to cope with an infestation on a few individual leaves or branches. With the gall mite, it's a different matter, however. These arachnids also cause abnormal growths on the leaf through their sucking action, producing deformities and sometimes reddish pimples for their nests. These make a cosy summer home for the larvae, which are less than 0.2 millimetres in size. They can be a considerable nuisance on blackberry bushes, though. The larvae happily take up residence in the unripe fruit: the infested area won't turn black in colour and is unfit for consumption. These berries are sadly best disposed of.

midday Sun on a summer's day is often sufficient to make their large leaves droop, even when they're in perfectly moist soil. They start to wither and die at even the first hint of dry soil. While lilacs, rhododendron and viburnum all have a stoic tolerance for heat, vegetables all need regular top-ups of water to keep them in good spirits.

Rising temperatures

The glaciers are melting, storms are becoming more frequent, and one dry year seems to come hot on the heels of the last. With temperatures ever rising, is there a huge environmental disaster waiting to happen?

What impact does climate change actually have? One thing is certain: it's not just a matter of things getting warmer. The combination of precipitation and temperature plays an important role in water management. More rain needs to fall than evaporates again, or else your garden will dry out like a desert. Of course, more water evaporates at high temperatures than at low temperatures. The warmer the water is, the more rain needs to fall to balance out the evaporation. If you live in a particularly arid area, such as the county of Norfolk, a rise in the average temperature by just two degrees could tip the balance. The warmer air allows more water to evaporate, so that if precipitation stays constant, there is suddenly a water shortage.

And that's not all. It has been suggested that the summer precipitation levels might drop, with more rain falling in the winter instead. So, even though the total precipitation quantity would be sufficient, the soil could be bone dry in summer, meaning that many plants would die. If there are desert conditions during the summer, it doesn't help the plants that it rains so much in winter that on average, theoretically, there should be sufficient water for the entire year. What makes the difference in such a context is the water storage capacity of your soil. Sandy soil, for example, holds hardly any water, while a high content of loess sediments releases moisture to your plants gradually over many weeks. So, you see, it is not easy to predict how your garden will fare. Only one thing is certain: it is definitely going to get warmer.

The impact on your garden

I don't want to go into the causes of climate change here, and certainly not to start apportioning blame. Because that makes no difference to your garden or its future. What's more important is the impact climate change has on plants and animals that have found a home in your garden. How will they cope with the changes?

To cut to the chase: you can easily help your plants to adjust. Because the closer your garden management style resembles natural conditions, the less the impact will be. Nature is well prepared for climate change. And that is hardly surprising on closer inspection. Our native tree species, such as the beech, for example, can live for over 400 years. This is a very long time on a human scale. Over the course of centuries, there are constant natural fluctuations in the climate. From the fifteenth to the nineteenth centuries, there were many extended periods when the temperatures fell steadily, so that harsh winters became the rule and glaciers spread. After these cold periods, which caused widespread famine, the climate warmed up again. This cyclical process continues to this day and is merely accelerated by the environmental disturbances.

Thus, within one or two generations of trees, there have been several profound changes to which beeches or oaks could not adapt. After all, adaptation works over many generations and then only very slowly: the deviations from the mother tree to the seedling are hardly noticeable. However, if a slight genetic change were only possible with the passing of each generation – that is, every few centuries – our native trees would have long since died out. The survival strategy of many long-lived plants is therefore not adaptation, but tolerance. The beech has an optimum climate in which it grows best: the average climate of Central Europe. However, it can also tolerate higher temperatures, lower temperatures and even deviations in precipitation. The result is an area of distribution that extends from Spain and Sicily to Sweden.

The general principle is that the older a plant can become, the more climate-tolerant it has to be. Consequently, all trees, but also many shrubs, need to be able to tolerate a broad spectrum of temperatures and precipitation levels.

Whether the trees and shrubs in your garden will be able to withstand the expected rise in temperature depends on whether they are currently living in their climatic comfort zone. By this I mean that their optimum conditions prevail in your locality, and that they can withstand fluctuations upwards and downwards both in terms of temperature as well as precipitation. This is true for most native species, but also for imports from other regions with similar temperate climates.

Some experts recommend choosing plants that are better equipped for a warmer climate in the face of rising temperatures. I think this is very unhelpful advice. Rising average temperatures mean that dry hot summers and damp winters without snow will become more frequent. But even in the future, winter will still bring hard frosts, only much less frequently than nowadays. In northern Europe, it's no use planting Chinese windmill palm, for example, which can tolerate the cold to an extent, but which will die at temperatures below minus 10°C. The only sensible approach to future-proofing your garden is to choose plants that are capable of weathering both summer and winter temperatures.

If you want to know precisely how climate change is affecting your garden, you should start taking a series of measurements. For this you will need an external thermometer and a rain gauge. This way you can record the two most important parameters and how they fluctuate over the course of the year. The thermometer should have a maximum and minimum indicator (it can be a mechanical one or electronic). From this you can make a note of the daily highs and lows in your garden. As the device should be installed in the shade and not directly against the wall of the house, I would recommend an electronic one, allowing you to read the values conveniently from inside. Over the years you will be able to compare the temperatures in your garden and to discern long-term fluctuations. By installing a rain gauge, you will similarly be able to measure the precipitation levels in your garden (electronic devices are also available, but they are more expensive).

If to buy both is too much of an outlay, I would at least measure the rainfall. Depending on the weather, you only need to check it every few days, and you will be monitoring the more important value. After all, while temperature deviations tend to be relatively small within a locality, the differences in rainfall can be quite considerable over a distance of just a few kilometres. Perhaps there's a hill nearby which acts as a meteorological divide that diverts the rain clouds from your garden. Or perhaps there's a small grove of trees which affects the air strata above. Where we live, for example, we get considerably fewer heavy showers than in our local small town, only ten kilometres away.

To gain some useful information about the temperature, with relatively little effort, it is enough to compare your garden's maximum and minimum values for a few days with those of the nearest public weather station. You will start to see that the deviations are very regular; for example, that it is always two degrees warmer in your garden. This means that you can start to apply the changes in temperature as measured by the weather station to your garden, adding on your extra two degrees, to stick with the same example. Together with your exact precipitation values, you will start to form a fairly accurate picture of how your local climate is changing.

The longer the period over which you take your measurements, the more valuable a picture you will build up. The weather can be capricious, after all, so the values of a single year may be insufficient to tell you anything meaningful. Only with an average of many years' data can you see the direction of travel and get a sense of what you and your garden need to adjust to.

To show you how fragile our water supplies are even in our supposedly rain-blessed climes, I would like to offer an observation from the reserve I manage. In various locations along the forest paths, I made some small ponds which were fed by water from nearby sources. It was a pleasure for many years to see how these small ponds became a haven for salamanders, as well as toads and frogs. They spawned and over the course of the summer I would observe the growth of this mighty little army of frogs (and knot of toads). But this thriving amphibian community is sadly a thing of the past. For Germany, 2003 was a record dry year, when there was no significant rainfall between March and October, and many aquatic biotopes completely dried out. And as the water disappeared, it meant a bitter end for our amphibious friends. The following years saw some particularly damp summers. Many a thunderstorm drenched our road and swept deep channels into the dirt roads; many a winter with thick carpets of snow made for saturated soil. And yet the groundwater has clearly still not recovered from the drought. Every year since 2003, the same thing happens: the ponds come and go, drying out every year, with the amphibian numbers dwindling along with the ever-dwindling water supply.

When I speak with colleagues, I hear of similar findings. The real calamity lies in the fact that disruption to the groundwater levels is not easy to detect. The rain gauge is useful only to assess the water supply from above (unless you have your own well). Strictly speaking, you can only determine the deficit of dry years compared to normal years and see if it rains in sufficient quantities in the subsequent years to balance out the deficit.

What are the consequences for your garden? For the vegetable patch and flowerbeds, a lack of rain is not critical at first. It affects only the upper layer of the soil, after all, and if this is too dry, you can simply water it. On the contrary, a dry season often means a particularly abundant vegetable harvest, because snails are kept at bay.

For your fruit trees and ornamental trees, it's quite a different matter. They grow deeper roots, to provide for themselves. So if, during a hot summer, you want to water your trees, remember that a very special problem can arise. In the section 'How to water properly' in Chapter 3 (see page 28), I mentioned that we should avoid mollycoddling our plants by watering them too regularly. Water them rarely, but when you do, really give them a thorough soaking. It takes a lot of water to moisten the roots of cucumbers or radishes, making sure you reach the deeper layers of soil to encourage them to grow deep roots. But to water a tree in a drought is a task of a completely different magnitude. To properly saturate the dried-out root bowl of an average mature apple tree takes around five cubic metres of water!

Now, of course, you could start sooner with smaller amounts of water, a few weeks after the last rainfall. But then the tree will begin to develop its roots more intensively in the uppermost soil layer. It will gradually become more dependent on your weekly dousing and will become less and less able to withstand drought. Because of its size, another very different hazard arises: the tree is at risk of becoming dangerously unstable. After all, a tree's roots are its anchor, grounding it and holding it steady in the autumn storms. If it gives up developing its deep roots, it becomes a very shaky prospect.

However, in case of extreme drought, it can make sense to intervene. The right time to take action is when the tree begins to discolour or even discard some of its foliage in the middle of summer. But remember: if you're going to do something, do it properly. You can genuinely help, but only if you give your tree a thorough drenching.

8

ASSESSING YOUR SOIL QUALITY

WHEN it comes to natural processes, only half of what's happening takes place above ground. Recent research has shown that bacteria and other primitive species reside up to ten kilometres below the surface. A millilitre of groundwater can contain some hundred thousand minuscule life forms, and it is possible that the total mass of the subterranean creatures exceeds that of all the animals and plants of the earth's surface.

Beneath the surface of your garden is a gigantic habitat, rich in species, which has yet to be properly explored. In fact, you consume some of it every day, in your tea or coffee. Countless minuscule organisms end up in your beverage along with the ground water, even after being treated before it reaches your tap. But since they are not harmful to us, and are often even beneficial, their presence is nothing to worry about.

For your garden and your plants, it is only the topsoil, the uppermost layer, which is important. The quality of this soil layer – whether it is rich or poor in nutrients, and whether it retains water well or is arid – essentially depends on the bedrock. After all, it is only the underlying rock that is present locally that crumbles into soil through the process of weathering.

Weathering occurs over extremely long periods. In prehistoric times, the landscape would have consisted initially of bare rock. Severe fluctuations in the temperature left cracks in the rock, which water could penetrate. Winter frosts made the water freeze, thereby expanding and shattering large lumps of rock into small fragments. Chemical and biological processes formed acids that broke down the rock further until it was decomposed to the finest substrate. Physical forces, such as stormy winds kicking up sand in the air, worked on the rock like sandpaper, reducing boulders and stones to dust. The result is a layer of soil of varying depths, depending on the altitude. It also includes a good amount of humus, which we will discuss shortly.

So, the weathering of rocks results in small particles, and the size and composition of these are crucial for the fertility of the soil. We tend to categorise soil types into sand, with large grains, silt which is finer, and clay which is finer yet. With pure sandy soils, where you can feel the grains of sand, the layers of soil are loose and water drains through easily. Silty soils are considerably more compact, but still permeable to water. Clay soils, on the other hand, consist of very fine particles. These soil types retain moisture and hold very little air. A combination of all three types is called loam. This ideal soil mix retains water and nutrients, it is well-aerated and, as long as there is sufficient humus, it makes for fertile growing conditions. But even with loam, there are differences in quality, depending on the rock type that forms the basis of the soil.

This process of soil formation continues to the present day, but it is happening much more slowly than in the prehistoric era. Exposed rock was defenceless in the face of the onslaught of weathering, while nowadays the Earth, with its sheltering layer of plants, is well protected. Indeed, this process is so slow that it can take several centuries for a centimetre of new soil to be formed.

Soil types

If you want to evaluate the quality of the soil in your garden, you need to know the underlying bedrock. Depending on the origin and age of the soil, it can contain a wide range of nutrients, which are released during decomposition and made available for plants to consume. If the earth was backfilled when the house was built or added later, then knowing your local bedrock won't necessarily help, as construction companies often use material sourced from mixed locations. You can easily check whether your garden was backfilled in this way, if the layer isn't too thick. If you ever need to dig a hole to make a foundation for a wall or a fence post, this is a good opportunity to take a closer look at the soil stratification. First you will usually find a darker layer at the top, which contains a lot of humus. You won't find this dark soil lower down, where the earth is still in its original state, because the humus from decomposed leaves and plants doesn't reach these greater depths. If the soil was backfilled at some point, this natural humus layer will be lower. This dark humus layer always shows the original surface. If you do several trial digs around the garden, you can find out whether your entire garden was backfilled or just certain areas.

If the soil in your garden is still largely unadulterated, it is worthwhile carrying out a more precise investigation. It isn't always easy to work out the geological source of your soil, but this only needs to be done once and is well worth the effort. If you are lucky, you won't even need to get your spade out. There are regions where the landscape has just one type of rock. You can find out whether this is true where you live by checking online, for example, on the homepage of your local geological society. These organisations publish regional maps that you can use to work out the type of rock forming the substrate in your area. The maps often also indicate local soil types and composition.

Another way is to seek out some old stone cottages in your locality. In centuries gone by, the material for house construction was whatever was available in close proximity, so you can tell at a glance which rock types are characteristic locally, just from the stone buildings nearby. This, however, only applies to the houses of ordinary folk, as the stone for more upmarket dwellings may have been transported several hundred miles. In some regions, there are very few stone cottages, because brick was the traditional building material. The reason is that vast tracts of land are made up of sandy soils where very little rock is available to quarry. Dark red houses built from loam or clay bricks are therefore usually an indicator of sandy soils.

If you want to work out precisely the nutrient levels of your soil, you can send off a sample for laboratory analysis. The best time to do this is in autumn or early spring, before you add any fertiliser. To do this, you will need to cut in with the spade as deeply as the roots of the plants in the patch of soil in question. A depth of 10 cm is sufficient for lawns, while vegetables may have roots down to over 30 cm, and fruit trees and bushes can have roots stretching down over half a metre. Dig up several spades full (from 10 to 15 different places), place these in a large bucket, mix the soil well with your hands and then fill a bag with approximately 250 to 500 grams. You will need to produce a separate sample for each bed and each different use. There are various organisations which analyse domestic soil samples, such as the Royal Horticultural Society.

And if this all seems a bit too complicated, there is another method which does not involve digging up samples or researching your neighbourhood's architectural history.

You can use so-called indicator plants to work out the nutrient levels and water retention of your soil. Each plant has an optimum range in which it is particularly competitive and can dominate over other species. For most species, you can look up the optimum soil conditions and use this information to work out what you are faced with in your garden. For this you need two things: a good classification book to help you identify species and a patch in your garden where you're happy to wait and see what grows, leaving it to nature's discretion. Fertilisers or other soil enhancers would distort the picture, so this set-aside area should have been left relatively untouched for at least a few years. It could be a lawn, providing you haven't reseeded recently and it has been left to grow whatever weeds thrive best.

Once you have set a corner aside to return to nature, you can start to keep a record of which species are most at home there. It is important to record as many indicator plants as you can, because just as one swallow doesn't make a summer, likewise one single plant doesn't prove a set of soil conditions. You can only form a coherent picture once you've spotted a few species. For example, white dead-nettles prefer moderately moist but nutrient-rich soils. If you identify several plants with similar preferences, such as sweet violet, you have a stronger body of evidence and can start to draw conclusions about your soil.

If, on the other hand, you find other species which seem to contradict the picture, such as heather, which is at home in acidic, nutrient-poor locations, there are two possible reasons: either the plants have been artificially seeded (and would not naturally settle there), or there are variations in the soil conditions that you have overlooked. This could be caused by, for example, a limestone gravel path. Lime raises the pH value, making the soil more fertile. Plants with a high nutrient requirement can grow at the edge of a limestone shingle path, while just a few metres away the flora suggests acidic conditions.

Indicator plants can also be used to assess the success of soil improvement measures. Every time you add fertiliser or dig in compost, this results in a change in the weeds that make themselves at home. Abstemious species like the oxeye daisy might be displaced by plants with greedier appetites, such as blackberry. If you observe an increase in stinging nettles or comfrey, then you might have slightly overdone it. You may see variations in the plant species after just one or two years.

But there is even more information to be gleaned from your garden's flora. The greater plantain, for example, is characteristic of compacted soil, such as at the sides of roads and footpaths. The diversity of wild plants also reflects the local climate. A case in point is the common foxglove, which prefers the Atlantic coastal climate to that of continental European; that is, a climate with mild winters and moderately warm summers.

And what is to be gained from all this detective work? If you know your gardening conditions, you can plan your planting with confidence and will be faced with fewer disappointments.

Encouraging humus

When animals and plants die, or they drop excrement, leaves or fruit, this organic matter is broken up by soil organisms as they eat and excrete. The brownish black matter that remains is called humus.

Humus contains an average of 60 per cent carbon, almost as much as lignite, or brown coal. This is also the reason for its dark brown, sometimes even black colouring. The carbon originates indirectly from the atmosphere. Plants take in carbon dioxide for photosynthesis: using sunlight as energy, they combine the carbon with water to create carbohydrates in the form of sugars and fibres. Animals that feed on plants incorporate this carbon into their own matter. When soil organisms, such as fungi or bacteria, digest and break down leaves, they exhale some of the carbon in the form of carbon dioxide (CO_2). A considerable percentage of the animal and vegetable tissue, however, remains in the soil in the form of humus. Everywhere that there is a permanent cover of green vegetation, e.g. meadows and the forest floor, more carbon accumulates on the ground than the soil-based organisms can decompose. This process is even taking place beneath your garden lawn. It is the first stage in the formation of coal, oil and gas. Fossil fuels are therefore nothing more than prehistoric humus. Of course, it is an extremely slow transformation, but green spaces represent CO_2 deposits and this is how they contribute to climate protection. A garden of 1,000 square metres (a quarter of an acre) can pack away up to one ton of CO_2 per year. This corresponds to the carbon dioxide emissions from a car journey of over 4,000 miles or a train journey of 15,000 miles (per passenger).

You can tell how well underway this carbon storing process is in your soil from a single spade full of soil. From the ground (turf) going down, the earth changes colour from dark to light. The carbon content in the soil corresponds with this colour spectrum, decreasing in quantity from top to bottom. The thickness of the dark layer also reflects how little or how much CO_2 is stored in the soil. Very little carbon is present below this upper layer, as can be seen in the paler colour. It is best to carry out such an investigation during a dry spell in summer, so that the colour of the earth is not affected by moisture, and the contrast between the layers will be easier to see.

The carbon storing process and the emergence of this dark soil layer is something that takes place only in permanent greenery or under trees. By contrast, open ground such as arable farmland, or neatly tended allotment beds, show quite a different picture. Here, the soil is left exposed over the winter and, in the summer, the Sun's rays directly warm up the clear ground that is left between the cucumbers, tomatoes and radishes. The warming effect prompts the soil-based microorganisms to gorge themselves in a feeding frenzy. The solar energy helps them operate at peak levels, meaning that they demolish the bulk of the humus layer within a few years. At first, this leads to explosive growth in crop plants, because the degradation releases a lot of nutrients. However, much of this nutrition washes down into the deeper layers of soil with the rain, meaning that much of the over-production goes to waste as plants can't process it fast enough. The growth will slow down, and to avert this the soil needs to have compost added to it every three or four years, or to be fertilised with manure on an annual basis, so that the worms and their soil-based friends don't starve.

Lime is a popular natural fertiliser, but you should be sparing with it, or avoid it altogether, as just like the warming rays of the Sun, it spurs soil organisms into life in the fast lane and means that the humus is devoured at top speed. This rapid reaction often frees up more nutrients than vegetable or ornamental plants can absorb, and the brown gold in your beds goes to waste. If the vast nutrient supplies are burnt off, a large part of the humus disappears, so the performance of the soil drops considerably below the value it had before being fertilised with lime. 'Lime makes rich fathers and poor sons,' as the saying goes, warning against an excessive reliance on the stuff in agriculture.

If you want to protect your soil's natural riches, the best way to shelter the humus layer is to leave a few harmless weeds between your crops. They provide shade and help keep the ground cool and moist. Rather than losing excess nutrients when they are washed into the depths of the soil, they are absorbed by this vegetation and can be recycled when these weeds are removed to the compost heap (or are dug straight back into the ground). Chickweed is one of the most suitable plants for this purpose. It spreads rampantly and quickly forms a dense, green blanket. It rarely threatens vegetables or perennials, and if necessary it can be pulled up without any great effort. Incidentally, it can also be put to good use in the kitchen: it goes well in a salad, or chopped and stirred into cream cheese, and is a traditional herbal remedy with various applications. At the start of the season, the bed can be turned over for planting, and after a few days, the chickweed will be up and running again in the form of seedlings.

Modern agriculture can't survive without humus either, and yet one often has the impression that the brown gold produced naturally in the earth has been forgotten about. All organic matter is removed – whether it's fruit, straw or chaff – leaving almost nothing behind for the soil-based organisms to munch on. The manure that farmers apply in winter is less appetising and also less beneficial for worms and fungi. Fertile nutrients are less likely to emerge from this concoction, which also often contains considerable amounts of antibiotic residues. While arable land does contain humus, the quantity is being steadily depleted. The soil that large-scale farmers benefit from today comes partly from ancient times. It is the remnants of old meadows, perhaps even ancient woodlands that once stood where the crop fields are today.

Useful garden residents

We have already discussed how humus is formed. But which soil-based organisms do we have to thank? There are thousands of species dedicated to the decomposition of organic waste.

The first stage of the nutrient recycling process is performed by animals. They gnaw and grind up leaves and stalks, digest them and excrete them, often with mucus. These herds of tireless garden helpers include earthworms, snails, mites, springtails and nematodes. Their output – crumbly, porous humus – retains water well and is crucial for soil fertility.

This humus, meanwhile, is needed by two other groups: fungi and bacteria. They could also get by without the preparatory work undertaken by their underground neighbours, but they are better able to consume and extract nutrients from the munched up, pre-digested material.

In terms of the number of individuals, the largest group is bacteria. One gram of garden soil can contain over 100 million microorganisms. There is almost no organic substance that they can't decompose, thereby ensuring that every living being returns to the cycle of nature after death. It is when these tiny organisms process the dead vegetable matter that carbon dioxide is released again from being stored in the plant's cells. In natural ecosystems, however, some of the humus reaches deeper soil layers where the living and working conditions for bacteria are less favourable. This preserves the humus and with it the carbon.

The last group of the soil residents is a bizarre one: fungi. These organisms belong to neither the plant nor the animal kingdom. In the past, I would not have hesitated before categorising them with plants, but the latest research reveals them to be closer to animals. Like animals, fungi do not photosynthesise, but rather feed on the organic matter of other organisms. In many species, the cell walls consist of chitin, like the exoskeleton of insects. The fruits of many species are particularly conspicuous – the umbrella-shaped mushroom or toadstool with a stem and cap – but these are no different in function to apples on an apple tree. Countless spores flutter from these fruits, to be transported away by the wind or by animals.

The actual functioning part of the fungus is toiling away out of sight below ground, with its thread-like filaments weaving through the upper layers of soil. One gram of earth can contain up to 100 metres of these minuscule filaments. Certain species, such as porcini or birch bolete, collude with trees to mutual benefit. The fungus weaves itself around the tree's roots and, like a ball of cotton wool, soaks up water and minerals from the soil, to pass on to the tree. In return, the tree excretes a sugar solution to nourish the fungus.

In the garden, a fungal colonisation can lead to some surprising sights, like finding a fairy ring of toadstools appear on your lawn. This structure is a charming side effect of the way a fungus grows. The main body of the fungus is the mycelium, the web of thread-like filaments, and as these inch their way through the soil they consume everything nutritious they encounter in their path. What is nutritious to a fungus is all kinds of dead organic matter, including the thatch layer of dead turf that builds up between the fresh grass and the roots. Below ground, the fungus spreads in every direction evenly, while at the centre the tissue dies off at the same rate after consuming the nutrients in the soil. Over the years, this results in a large ring structure.

If the fungus fruits in the autumn, it can only do so over tissue that is still living, which means the mushroom caps pop up in a circle on your lawn or among the leaves on the forest floor.

Fungus can sometimes lead to discolouring of a lawn. Where the mycelium is currently alive and traversing through the soil, the grass above is often considerably darker and stronger than in areas without colonisation. This leads to the misconception among gardeners that the fungus weakens the grass. In fact, the opposite is the case: the decomposition of dead matter creates humus which replenishes the lawn's nutrient supply. The dark colour of the grass over the mycelium shows how healthy the grass is.

However, a fungal colonisation does have its downside: in some species, the dense network of filaments can clog up the soil, preventing rain from seeping through to the grass roots, and thereby resulting in very parched turf in summery dry spells. However, this disadvantage is outweighed by the enhancement to soil quality, meaning that fungicidal measures are rather counterproductive.

Soil compaction and its long-term impact

Across Europe, the earth is no longer in its natural state. Before being settled by humans, the landscape was dense with primeval forests. The closed, dense tree cover was the best possible protection for the fine, loose soil, and all processes took place at a slow and moderate pace under the canopy of beech, oak or ash. Humans removed this protective layer around their growing settlements by clearing vast tracts of woodland. But this is not all: the early farmers left an indelible mark on the soil when their oxen pulled wooden ploughs, dragging the topsoil into ridges and furrows. These ploughs turned over a very shallow layer of soil, no more than 20 cm. The soil below this was smeared by the plough, resulting in a clogged-up layer called the plough sole, blocking the pores in the earth and stopping air and water from seeping through. This effectively suffocated the soil life beneath this layer and meant water could not be fully absorbed after heavy rain. The result was a bathtub effect: after rainfall everything was submerged, whereas in dry periods no moisture could be drawn up from below.

Shepherds and goatherds have also wreaked havoc with their livestock over the ages. The surface of the ground has been beaten down by the animals' hooves, causing further damage to the pores through successive layers all the way to the surface.

As a student, I took part in a field trip to the Swabian Jura, a mountain region in south-west Germany. The local forester showed us a soil profile in the woods; that is, a dug-out patch revealing the individual layers in the soil. It was clear to see that a flock of sheep had lived here about 300 years ago. The crushed upper layer of soil had still not recovered.

Even today, modern farming continues to exert a massive pressure on the soil. Consider our bulky agricultural machinery and equipment: in comparison, the cattle-drawn ploughs were flyweights. Even in forestry, men and horses have been replaced by huge, heavy harvesters weighing up to 50 tons. It's no wonder that the ground is running out of breath.

Why am I telling you this? This cultivation and degradation of the land, whether it was carried out in the Stone Age or 20 years ago, has an impact on the soil to this day. Such interventions leave a lasting mark which, like a memory, it is not easy to erase. The chances are high that your garden, which is likely to be on former agricultural land, is also marked by such an experience. Only an uninterrupted woodland history would guarantee an intact pore volume in the subsoil. There will be nothing of the kind in modern construction sites, where soil is shoved around with heavyweight excavators on caterpillar tracks.

But don't worry, this doesn't mean that your soil is worthless. Depending on the level of compaction, it will be restricted in its effectiveness, and it is worth being aware at what depth the damage begins. The impaired pore structure is found below the first 20 cm. Paradoxically, the top layer is usually fine, although it has borne the weight of heavy machinery. The reason why this layer is more porous is the frost. When the ground freezes, together with the recent rainwater it contains, it is only the upper 10 to 20 centimetres that freezes. As the ice expands, it breaks up compacted soil, opening up cavities and leaving the topsoil better aerated. The soil-based life forms can return and become active again, at least in this upper layer. Moles and voles also help ventilate the soil with their underground corridors.

There are a few ways to recognise a compacted layer. On the one hand, there is low water permeability. When it rains a lot, the ground becomes waterlogged, since the water cannot seep down. This can initially give the impression that it is a particularly damp location. However, in contrast to this, compacted soils also dry out again very quickly, particularly on hot summer days when the moisture is urgently needed. Seasonally humid soil like this poses a particular challenge for vegetation: species with a high water requirement become dehydrated in summer, while less thirsty plants can drown in the wet spells.

Another clue is the penetration resistance when you're hoeing, for example. If you want to loosen up the soil any deeper than the upper 20 centimetres, which is always being regenerated, you will have to use considerably more force if the soil is compacted. You'll know for sure if you take a closer look: dig up a sample to a depth of about 30 to 40 centimetres, and compare it with the soil from the surface. While the latter is fine-grained and crumbly, the compacted soil from lower down will contain very few pores. It is more like a lump of clay in texture, and when dried it breaks into small, angular chunks. You don't get this with porous, uncompacted soils. The deficiency of oxygen, caused by the lack of pores, is visible in the light greyish colouring, dotted by small, brown rust patches from iron in the soil. This may have occurred naturally, but in most cases this kind of degraded soil quality is caused by mankind's agricultural activity.

The consequences of compaction are not limited to drainage and water management. Most plants cannot cope with an oxygen deficiency in the root zone. The delicate runners suffocate as soon as they reach this oxygen-starved level in the soil. This is bad for root vegetables, which grow into wonky, forked shapes and cause nothing but irritation for whoever comes to peel them in the kitchen.

In the case of trees, the frustration is on quite another scale: a tree will struggle to root properly in compacted soils and thus won't be safely anchored. The problem of shallow roots, often associated with spruce trees, has its origin in poorly aerated soil. From a statistical point of view, this gets a bit risky for coniferous trees when they reach heights of over 25 metres. They are exposed to the risk of toppling, because they retain their foliage even in winter – the stormy season in Europe – and therefore offer a full sail for the wind to attack, in contrast to deciduous trees where the wind can rush through the bare branches. A conifer can be uprooted at wind force 10, especially following rainfall which can't drain into the compacted earth. A root system that reaches only 20 cm deep can't keep stable in the resulting muddy, gloopy soil, making the tree prone to toppling.

If the conditions in your garden suggest the soil is in this kind of state, there are still many ways to raise healthy, happy trees. Firstly, there are two species of tree that can help break up the compacted soil areas. Oak and fir trees at least partly eliminate the sins of the past, since their roots stretch deep into the soil no matter what the oxygen levels are like. Besides these, deciduous trees should generally be safe, since dropping their leaves in winter makes them stable even as large trees. Species which remain relatively small, for instance fruit trees, are a particularly safe option.

Once we are aware of the lasting impact of soil compaction, we start to see burrowing animals in a whole new light. We should recognise field voles and common voles, and above all moles, for the benefits they bring. While voles tend to burrow at relatively shallow depths and rarely descend lower than 50 cm, the mole tunnels down twice as deep. Their burrowing depends not so much on the hardness of the soil as on the supply of food: the more earthworms and grubs there are, the more merrily these blind subterranean mammals will multiply. So don't get upset by the brown molehills on your lawn: instead, see it as the beginning of your soil's recovery, because the tunnelling activity supplies aeration to the deeper layers of soil. Soil life can breathe again, and the garden increases in fertility. Incidentally, you can recycle the fine-grained soil from the molehills in your raised beds and pots, or you can simply spread it around over the grass. Within a few weeks, there will be very little left to see.

Voles are less likely to have such a positive impact as they burrow much closer to the surface. Plus, they tend to help themselves to your vegetables and flowers, so overall the scales tip in favour of the mole.

Preventing erosion

You will perhaps now have a good idea which type of soil you have, how slowly it forms, and how precious it is. You can influence the humus component to a certain extent with fertiliser or by mulching with compost. The actual soil, however – that is, the sedimentary loess particles or clay minerals – is not something you can simply reproduce. Of course, you can pour on topsoil, but this is a somewhat aggressive intervention considering that the soil life beneath this backfilled area will be largely destroyed. If you can, it's best to make do with the soil that is already available in your garden. And yet, in many cases, the result of erosion is that the quantity of soil becomes less and less.

The ideal conditions, with the lowest rate of erosion, are found in the forest. Beneath the canopy of the trees, the soil loss per square metre per year is less than one gram, i.e. less than the amount that is created in most cases. In forests and woodlands, the soil layer is constantly growing thicker.

Crop fields represent the other extreme. Out in the open, wind and water can cause the run-off of up to 10 kilograms of soil per square metre per year. It's not a matter of a one-off landslide, but the sum total of what disappears year by year. And as the regeneration of soil is such a slow process, it has a greater and greater impact on fertility.

Erosion is not a regular occurrence, but something that happens from time to time, in extreme weather events. The landscape is riddled with ditches or channels, like empty streams. When there is a heavy rainstorm, or when heavy snow cover melts in the spring, the soil cannot absorb the water quickly enough. Instead, much of the water runs off the surface, concentrating in these channels. This is a very visible form of soil erosion, with the result that these grooves in the earth are deepened a bit more each time.

Modern agriculture leaves a lot of artificial depressions in the bare fields in winter through the act of ploughing, and the rain can very quickly wash away the earth along these channels.

Erosion like this can also happen in your garden. Have a look and see whether small rivulets form in your beds when it is raining heavily. As soon as the water turns brown, you can see your beautiful soil washing down the drain. A permanent cover of vegetation can help, for example, so it might help to sow a grass or cereal crop to grow over winter, after you have harvested your vegetables.

Weeding is another form of erosion. The roots of those unwelcome plants embed themselves deeply among the desirable ones, so that when you yank them out, some soil inevitably clings to the roots. Shake each plant thoroughly before throwing it on the compost heap. For even if you leave just a few grams of soil on the small root ball, this could add up to kilograms over the course of the summer.

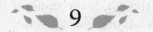

9

NATIVE FLORA AND EXOTIC GUESTS

W HAT is it about a garden that is so relaxing? I think part of it is that there's something reassuring in the fact that our plants can't simply get up and walk off without us. Imagine what it would be like if your tomatoes, roses or magnolias were mobile: you'd have to build fences around every bed to keep them in, or resign yourself to letting them escape. But thankfully, once our shrubs and perennials are planted, they stay pinned to the spot. But wait a minute: can we really say that a plant spends its entire life in just one place? Far from it. Not every species of plant is quite so well-behaved.

First, though, I'd like to talk about something that we tend to take for granted: leaf colour. Given the choice, plants would probably deck their leaves out in every shade under the Sun, resulting in a veritable smorgasbord of colour. There's one colour, however, that wouldn't make the cut, and that colour is green.

Green leaves and variegation

You may remember learning at school that plants use sunlight to produce carbohydrates – sugar, cellulose and other structural components and nutrients – from carbon dioxide and water. It goes without saying that photosynthesis isn't something we can directly observe. One thing you can see very clearly, though, is the result of the process, as the by-product of photosynthesis is green.

Chlorophyll, a magnesium-containing hydrocarbon, is one of the substances that enables photosynthesis to take place. All leaves contain this green substance, which helps plants exploit light energy. And why is it green? Light is composed of many different wavelengths on a spectrum from ultraviolet to infrared. Not all of these wavelengths are used during photosynthesis, however, so any light that the plant doesn't need is reflected in its leaves. Green is the colour that is filtered out, the electromagnetic waste product. The green wavelength is all that remains of the Sun's rays after the other wavelengths have been absorbed. If plants were able to take in the full spectrum of light, they would look black.

The green you see in meadows, woodlands and gardens is evidence of countless plants at work. With the exception of the unique life forms living in the depths of our oceans, our planet is home to three forms of life: plants, which feed on sunlight, animals (including humans) and fungi. So when you gaze upon the green grass and foliage in your garden, you're also looking at the motors that power all life on Earth.

Sometimes, our gardens are also home to plants unwilling to restrict themselves to the various shades of green. Red-leaved specimens such as Japanese maple, purple leaf plum or copper beech, for instance, are cultivated varieties of the original green-leafed variety. A considerable number of ornamental plants are also variegated, i.e. have multi-coloured leaves. These genetic mutations would put them at a disadvantage in the wild as their discolouration is the result of low levels of chlorophyll. This relative lack of green pigment means that even in summer the leaf's colour is influenced by red carotenoids, pigments normally only seen in autumn.

In the wild, these impoverished plants would be outpaced by their healthier neighbours as they're restricted in the quantity of sugar they can produce and their growth is therefore stalled. Once they fall behind in the race, they're done for: slow coaches don't get as much light as their competitors and die before they even have the chance to reproduce.

Of course, a garden is not a level playing field. In this domain, humans decide who wins and who loses, which is why red- and purple-leaved varieties are only found in cultivated landscapes. These plants also come with one great advantage for small gardens: there's no risk of them ever outgrowing their welcome.

Trees and shrubs: friends or foes?

Trees can be more violent than you'd think. These giants, with their leafy boughs that sway serenely in the summer breeze, can in fact be quite unpleasant to other species. Their primary weapon is their trunk, which can – in extreme cases – grow to a height of over 100 metres. But if a trunk can be a weapon, does this mean that plants can actually do battle with one another?

It's a silent struggle fought in our backyards, and the prize urging the victor onwards is light. Sunlight is in limited supply and every year, hundreds of thousands of seedlings pit themselves against one another to be in with a chance of snapping it up. The only way a plant can win this battle is by securing the pole position. Pulling clear of its competitors, growing taller and broader, the winner leaves the others in the gloomy shade. The losers become stunted and sickly until finally they starve in the half-light. This is why most plant seedlings return to humus. The benefits of a trunk now become clear: the trunk is what helps trees – be it a cherry, spruce or oak – reach otherwise unattainable heights. Grasses, shrubs and perennials don't get a look in. Left in the shade, they simply wither away, which is why the floors of particularly dense forests are so barren and empty of plant life.

A garden is a small biotope for a tree, and such bullish behaviour can clash with a gardener's desire to cultivate as wide a variety of plants as possible. Despite their belligerence, however, there's no reason to shun these arboreal giants completely: oak and birch, for example, are much more compliant and let more light through to the ground than other trees. Fruit trees also make good neighbours to other garden plants; their height is kept in check by the process of grafting onto relatively short trunks.

Few garden owners dream of planting an entire forest, however; after all, we tend to only feel truly comfortable in open spaces. The Sun's warmth should be allowed to reach the flowerbeds and the patio, as no one wants to be constantly in the shade. Technically speaking, domestic gardens are effectively steppes, grassland habitats dominated by low-growing plants and just the occasional tree, usually positioned at the edges of the plot.

Shrubs are often a better choice: they seldom grow taller than three metres and can be used as a screen or partition, while still letting enough sunlight through. Shrubs are also very forgiving of heavy pruning and grow back with real gusto, something that is no surprise when you consider the many large grazers they have to contend with in the wild. Gazelles, wild cattle and deer nibble at their foliage, forcing them back to square one on a regular basis. Whether this pruning comes courtesy of a wild animal or a gardener's shears is of little consequence to these woody perennials: they're used to being kept in their place, which makes them a perfect addition to your garden.

Invasive species

Species have migrated ever since life has existed on Earth. Whether it's to escape climate change or to access new habitats, one thing is absolutely certain: no plant or animal stays in the same territory forever.

Admittedly, such changes can take hundreds – if not thousands – of years. What we're witnessing now, however, is a rapid acceleration of this migratory roulette. Transported in humans' cargo and luggage, whether intentionally or not, ever more plants and beasts are finding their way to new continents. Neophytes – that is, botanically speaking, introduced flora – have had a pronounced effect on our landscape, as the following examples show.

Over half of the total area of Central Europe is used for agriculture, mostly in the form of arable fields. These swathes of countryside were originally covered with dense forest, which was felled for the land to be ploughed, and most of the species now populating our countryside come from other continents. Potatoes, maize, peppers – we've become so used to these newcomers that they no longer stand out as anything unusual. Our woodland has also undergone a similar transformation many times over: regimented lines of spruce and pine are now found in places where beech and oak once rustled in the wind. Species change has taken place across almost three-quarters of Central Europe and this change has been brought about quite deliberately. Nature has nothing to do with it anymore; we now live in one of the most densely populated regions in the world, placing the emphasis on a constant supply of food and consumer goods. Take a look in your garden sometime and guess how many of your vegetables, flowers and trees are actually native to your region. I don't mean to wield the moral club here or suggest that native species are necessarily superior.

An organic garden with imported cultivars can still provide a safe habitat for threatened native species, while making a significant contribution to our food supplies, not to mention our wellbeing. The point is that, for a very long time, most of the plants we see around us have been neophytes. Most of them manage to stay out of the headlines because they behave cordially. They stay out of mischief by sticking to the patch they've been allocated, restrain themselves from spreading uncontrollably, and disappear as soon as we stop maintaining them.

Plant species only crop up in the media when they don't respect this bargain, breaking free of gardens and fields to run amok through the countryside. We'll take a closer look at some of these troublesome species shortly.

Like animals, plants can also emigrate. Not on four legs, granted, but their embryos are very capable of changing their location in the form of seeds or fruit. These seeds and fruits also have to go on a journey to find a new home where they can grow into a sexually mature plant, produce seeds, and so on. Light, fluffy seeds are carried on the wind many hundreds of miles, reaching new shores and then descending to the ground. Heavier seeds, such as nuts, need a little helping hand from animals and are collected and stashed away by squirrels, mice and jays. During the hungry winter months, these hoarders go back to look for their hidden provisions. But their memories sometimes fail them and a proportion of their haul escapes unscathed, going on to germinate the following spring. The speed of travel for nut trees, oak and birch is very slow at a couple of kilometres per decade, but that said, it's also rare for many of their animal helpers to emigrate so far afield.

The speed of plant travel has now reached a whole new level, however, and we humans are the facilitators. Whether by car, train, ship or plane, plants now emigrate using modern transportation just like we do, so it's no surprise that so many have succeeded in making the leap from one continent to another. We can see just how much their success has changed and redesigned our landscape when we consider our agriculture, and in this new look, native species have become a minority.

Most agricultural crops have one thing in common: they stay in their allocated plots. When have you ever heard of an out-of-control potato or an unruly cabbage? The answer is never, as these immigrants aren't particularly competitive; they're not even particularly resistant to frost, and survive only because of the TLC we lavish on them. If humans were to withdraw their care, most cultivated plants would vanish again without a trace. The vast majority of agricultural crops fall into this untroublesome category.

There is, however, another group of migrants that doesn't stay within its designated preserve – be it a field or your garden – and which absconds at the first opportunity. Their seeds are spread by wind, water or bird, and they are quick to colonise entire tracts of countryside. This colonisation occurs at the expense of indigenous species, with the significant consequence that native varieties are displaced.

Often, the starting shot for such an invasion is fired in fashionably landscaped gardens. This is how, at the beginning of the nineteenth century, Himalayan balsam (*Impatiens glandulifera*) was introduced to Europe from Asia. With shoots reaching over two metres in height, its pink flowers made the plant an attractive addition to flowerbeds. In autumn, it produces several thousand seeds, after which it dies back. The name *impatiens* ('impatient') refers to the explosive mechanism that enables the plant to scatter its seeds up to several metres away. If there happens to be a stream or river nearby, a few of the seeds might fall into the water to be carried away elsewhere. Entire riverbanks are then rapidly colonised and native plants are more or less smothered to death. A thick carpet of leaves then develops, suffocating every form of plant life, and since Himalayan balsam dies back in autumn, the banks of rivers and streams are left with no protection, allowing the bare soil to be swept away by the winter rain.

Giant hogweed and Japanese knotweed are another two aggressive garden absconders.

Yet garden centres still persist in selling plants that we would be much better off excluding from our gardens. A few that spring immediately to mind are highbush blueberries, which originally come from North America and feel right at home in our European countryside; buddleia, a plant that can completely take over fallow land; and golden plume (aka Canadian goldenrod), which spreads along river banks.

So what steps can you take to make sure you play no part in this invasive species roulette? One safe bet is to ensure you always choose native species when planting: although they're still quite capable of becoming escapees themselves, they're usually more than happy to stay within the boundaries set by your garden fence.

A lot of the garden plants available to buy are, however, exotic imports. People are always drawn to the unusual, and no matter how abundantly a native plant may flower, shoppers will often overlook them in favour of drabber options just because they're new to the scene. The potato is a good example of this. When it first arrived in Europe – and before its value as a food crop had been discovered – the potato was sold as an ornamental plant. When you get the chance, take a look at the small white flowers of a potato plant in bloom: nothing particularly exciting. Once the potato started to be cultivated *en masse*, it became less unusual and lost its allure, so today, it would scarcely cross anyone's mind to plant this tuber in a flowerbed.

Of course, that isn't to say you should forsake non-native species entirely. It does make sense, though, to go online and find out about the plants and how they behave before you buy. There are many websites that provide up-to-date information on invasive neophytes. As you'll see, most plants are very tame, staying in their allotted flowerbeds and not causing any trouble. The number of plants that actually escape from gardens is very low. The 'one of ten' rule works as a rule of thumb: only one out of every ten introduced species can survive in our European gardens. Out of ten masters of survival, only one will succeed in making the leap over the garden fence and into the open countryside, where it will then be encountered in isolated outcrops. The only real problem cases are the 10 per cent of this group that then starts to displace native species after their escape. Ultimately, this means that it's just one plant species in 1,000 that should be left well alone, and this is a restriction we can live with.

Uninvited guests can also arrive in our gardens as stowaways in products such as bird food. One such guest is ragweed (*Ambrosia artemisiifolia*), a plant that can turn a garden idyll into absolute hell for hay fever sufferers. According to the Bavarian State Research Centre for Agriculture, one single ragweed plant produces up to one billion grains of pollen, which have a much higher allergy potential than grass. Since ragweed is a close relative of the sunflower, it can't be weeded out from sunflower fields, meaning that its seeds often end up in bird food. These seeds are then left behind in your garden and can germinate at any time over a 40-year period. If feeding the birds in winter, look out for the words 'free from ragweed seed' or similar printed on the bird-food label. This offers some protection for the future at least; it remains to be seen how much damage will be wrought by the bird-food time bombs of decades past.

As a species, ragweed is a weak competitor, and only really stands a chance in clear soil. If your garden is full of plants, there'll be no free gaps in the flowerbeds or on the lawn for it to gain a foothold.

10

FUR AND FEATHERS

As far as plants are concerned, we are mostly in control of choosing which ones to host in our gardens. Animals, on the other hand, are able to come and go as they please, often helping themselves to our plants and shaping our gardens in the process. This outside intervention means our garden plants may develop in ways that we hadn't entirely foreseen.

Territorial behaviour

When we humans set up boundaries around our land, we're actually behaving exactly like animals. While a dog lifts its leg to mark out its territory, we plant a hedge or erect a fence. The fence sends an unmistakable warning to outsiders: this is occupied territory and unauthorised entry will not be tolerated. Unlike our fellow creatures, however, we mostly respond to transgressions of these boundaries with lawsuits these days rather than physical attacks. The key difference here is our transferral of the threat of force, but in biological terms, we all march to the same tune. Another feature of the animal kingdom is that territorial boundaries are only ever observed by members of the same species. Animals from outside your species don't recognise this land as belonging to you or as your property. Take the neighbour's cat, for instance. No matter where the boundaries of its mistress's property begin and end, this moggy has to settle the score with its rivals and all too often, its de facto territory extends no further than the front door.

Have a closer look at your flowerbeds and right away you'll cross many different territories staked out by many different animals. Your presence doesn't disturb them in the slightest and in fact most will hardly notice you, simply because they don't see humans as rivals. Birds, mammals and insects aren't interested in our impact on their territory either. I'd like to use one or two examples to demonstrate just how many overlapping territories may exist in your garden at any one time.

Most songbirds require at least one hectare ($10,000 \text{ m}^2$) of space per breeding pair. The bigger the species and the more specific their feeding needs, the further the boundaries of their territory will extend. The great spotted woodpecker, for instance, feeds on ants and the insects that live in dead wood, and requires a territory of 30 hectares. The reclusive black stork, meanwhile, preys on a territory of over 100 km^2. Your garden, therefore, forms a mere fraction of a bird's overall stomping ground.

Of the mammals, mice have the smallest territories, coming in at an area of 10 m^2. Squirrels' territories measure several hectares and a fox needs a territory of at least 20 hectares for itself and its offspring. With domestic gardens rarely exceeding a few hundred square metres, the territorial needs of humans are relatively modest in comparison.

Human and animal territories differ in one major sense, however: while our fences remain a valid boundary every day of the year, most animal species actively mark out a territory only when they're breeding or rearing young. When autumn comes and their offspring can fend for themselves – or have become somewhat more independent, at least – the territories disappear. This is also the reason why single species are seen in larger concentrations at this time. Peace generally reigns among the birds perching in the bushes and trees around the bird feeder at this time of year, although when food is at stake, a few minor scuffles are always a possibility.

The marten's preferred method of marking its territory can have far-reaching consequences for those affected. These mammals signal their hegemony over an area by spraying it with a fluid secreted from their anal glands. In the area where I live, the bonnet of a car parked in a driveway is one location often marked in this way. The residual heat of the switched-off engine makes for a warm, dry hiding place: what could possibly be a more cosy den? I have found that provided you always park your vehicle in the same place, not a lot will happen; the new resident may drag in some comfy padding – a little soft furnishing, if you like – and every now and again, you might come across a few shreds of mouse, remnants of the marten's dinner. Running the car won't interfere with any of this activity, but should you decide to stay overnight at a friend's or relative's, and you park your car in front of their house, that's a different matter entirely. This territory might well belong to a different marten, and the only scent that should be present in this territory is his own. When you park your car, you introduce the scent of a rival, a rival that has had the audacity to leave its own mark. This perceived invasion makes the marten see red, prompting an attempt to eradicate the interloper's scent with an aggressive physical response. Rubber tubing is a particular target for these attacks, and an incensed animal will sometimes claw and bite at the bumper. If vital components are destroyed in the process, the engine can fail completely by way of an 'act of nature' that generally isn't covered by insurance.

If your car gets mauled in this way, it's almost always due to an overnight stay. And that's not all: the marten that lives near your friend or relative will mark your car with its own scent signature too, sending your own resident marten into a corresponding rage. The only remedy is to wash the engine and take defensive measures. I've tried everything: I filled the engine compartment with toilet rim blocks and bags full of dog hair, placed chicken wire under it and pepper on the motor itself, but none of that offered protection for very long. When our last car was butchered in the same way, we finally resorted to installing an electrical device in the engine, with small metal plates fitted at the marten's usual routes into the bonnet, all professionally wired by a mechanic. From that point on, everything calmed down. The small metal plates work like an electric fence, giving the animal a mild shock when it touches the metal. After that, it avoids the hostile area and focuses on commandeering the rest of the garden instead.

Most of the animals in our gardens are really very tiny, and their territories equally so. Whether mouse, insect or spider, these creatures and their diminutive territories barely attract our attention. When they start to interfere with our garden plants, however, it's quite a different matter.

Those that help and those that harm

Heinz Erven was a German pioneer of organic gardening. As a young man, I once visited his hilltop garden overlooking the Rhine, which he described in his book *My Paradise*. The eccentric Mr Erven's rigorous approach to organic gardening was revolutionary at the time and made a lasting impression on me. To this day, I still remember the luscious, rampant vegetation, the neatly trimmed elderberry bushes and the upside-down flowerpots filled with straw, hanging all over the garden. These were houses for earwigs which, Erven claimed, killed off any pests in his crops. Back home, I was full of inspiration and eager to get going, so I hung up straw-filled flowerpots of my own all over my parents' garden.

PREDATORY MAGPIES

One day last spring, I was looking out of my office
window into the garden, where a battle was raging. A magpie had
taken a starling chick out of its nest in the old birch tree and was
attempting to peck it to death. My wife ran to the door and chased
the magpie away. The little starling's head was bleeding but aside
from that, it was still looking sprightly. I took a ladder from the shed
and put the little bird back inside its nest along with its sibling, and a
short while later, the parents came back to feed the pair.

But were we right to do what we did? From the magpie's point of
view, definitely not: it had found food for its young, who were waiting
hungrily back at the nest for its return, and we'd deprived them of this
succulent morsel. Besides, isn't it for this very reason that most birds
raise two broods instead of one?

After all, do we feel any sympathy for the offspring of the butterfly,
the caterpillars so mercilessly preyed on by tits and redstarts? What
about the mother mice that get devoured by cats or owls and whose
offspring then wait in vain for their return?

I know that we feel a particular sympathy towards young birds,
and I certainly won't stop intervening when I see such dramas unfold
in future. Nevertheless, it's not right to interfere in this manner, as
doing so implies that the magpies are doing something wrong when
all they're doing is following their instinct to eat or feed their young.

Magpies become troublesome when they start to proliferate close
to humans. This only occurs because our actions have changed the
ecosystem so much that it has become a veritable magpie paradise:
magpies in large numbers, therefore, only highlight our own mistakes.

Furthermore, their bad reputation means that we frequently
overlook the beauty of these intelligent creatures. But imagine for a
moment if magpies were threatened with extinction: what joy would
we feel then, when we caught a rare glimpse of this bird and its
striking markings?

I now know that my enthusiasm back then was a little excessive. While earwigs do indeed eat aphids, they also do plenty of their own damage to vegetable plants and love nothing more than riddling leaves with holes. In general, dividing the animal kingdom into beneficial animals and pests suggests that such clearly defined categories exist in nature. Nature isn't as straightforward as that: our fellow creatures pay little or no heed to the goals of the human species. And when we actively try to suppress certain species, it becomes even clearer just how little we understand of the synergies of the living world. A prime example of this comes from the early twentieth century when cane toads were introduced to Australia. Their intended deployment was pest control on sugarcane plantations; they were invited to devour the beetles that were demolishing these sweet stems of sugar. Upon release, however, the toads rejected their allotted task and started to interfere with the native wildlife instead. The secretions released by the toad's glands are lethally poisonous to amphibian-eating animals, such as goannas and snakes, but these animals had no way of knowing that. And so, the cane toad has continued its advance from the sugarcane plantations of north-eastern Australia to the west, seriously depleting native species, with no end in sight.

Modern methods of genetic engineering can also be used to combat pests directly: genes providing resistance to certain diseases, for example, can be inserted into laboratory animals that are then released to mate with wild members of their species. We can only hope that this doesn't become established practice: every animal is part of a food chain that can disintegrate entirely if just one member is removed. The aphid, a creature we consider a nuisance, is no exception to this: their sugary secretions provide sustenance to many insects, including ants and bees.

The categorisation of animals into good and bad, beneficial organisms and harmful pests, is woefully simplistic and fails to take into account the complex relationships in nature. Rather than controlling an undesirable species and encouraging a desirable one, it's much more sensible to reinforce the ecological balance in your garden. The niches offered by flowerbeds, shrubs, piles of dead wood and trees are a source of great biodiversity, and where diversity reigns, no one species can become out of control.

So, we should take care to ensure that nuisance species aren't allowed to disappear entirely. But can we at least keep them in check when they start to inconvenience us? As we'll see, this too is more easily said than done.

Predators and prey

If your garden is designed to attract beneficial animals, then logic dictates that the pests infesting your vegetable patch should decline in numbers. Whether your aim is to attract slug-hunting hedgehogs, aphid-munching ladybirds or caterpillar-guzzling tits, if you offer them a pleasant biotope then garden pests will be a thing of the past. I've been familiar with this philosophy since I was a young man, indebted, once again, to Heinz Erven.

In reality, the predator–prey ratio is a little more complicated than that and has very little to do with our desired outcome. The link between predator and prey is best illustrated by a study in Canada, in which researchers investigated the interaction between moose and wolves. Granted, you're never likely to find these large mammals in your garden, but the same principles underlie all predator–prey relationships.

There's an island in the Great Lakes region on the US–Canada border that was home to a population of moose. Much to the dismay of the foresters, these moose feasted on young tree saplings to such a degree that hardly any remained. During a particularly harsh winter, the ice on the lake stretched so far that a pack of wolves was able to reach the island. Upon encountering the moose, the wolves made short shrift of them. Thanks to this abundant source of meat, the numbers of these grey hunters increased rapidly and it became harder and harder for the moose to escape. The herbivore population declined and the young forest flourished once more. All of this was in stark contrast to the wolf population, which was falling steeply: hunting the last remaining moose was becoming ever more arduous and many wolves were starving to death. As a result – and to the detriment of the young trees again – the moose population recovered and within a few years, the wolf population increased too. This game kept repeating itself until the moose population reached a particular low and the wolves starved.

The bottom line is that predator and prey populations influence one another, and ebb and flow in waves that crest at different times. Each cycle starts with a sharp increase in the amount of prey and once this food supply increases, the predator population also experiences a surge, creating the two opposing waves shown in the example.

But enough theory for now. What does any of this have to do with your garden? In your backyard, the aphids, slugs and caterpillars are the prey and the ladybirds, hedgehogs and tits are the predators. The beneficial animals (the ladybirds, hedgehogs and tits) need a huge population of pests to consume in order to multiply; otherwise, their offspring will have nothing eat. Your garden is a good place for watching this cycle in action: first of all, a wave of aphids and caterpillars will sweep through your vegetable patch and flowerbeds. Then, in summer, the ladybirds will finally arrive. Afterwards, tits (and other beneficiaries of this bountiful supply of insects) will raise three broods instead of two, so that come autumn, they are present in greater than usual numbers. This means that the population of beneficial animals is always at its height when it's too late to be of any help to your garden: their numbers merely indicate the extent of the aphid infestation in the surrounding landscape.

In the year 2009, for example, the Baltic coast experienced a massive ladybird plague. Holidaymakers, who had hoped to spend their vacation lying on a sunny beach, were hounded by the pests and in some cases forced to flee the coast entirely.

In many cases, though, it can be helpful to provide shelter in your garden for tits, hedgehogs and other species. These species are often unable to find a suitable biotope in land cleared for cultivation, where the natural equilibrium is lost, despite an abundance of prey. A nest box here and a pile of brushwood there can offer a solution. And if nobody moves into one of the nest boxes or piles of sticks, take comfort in the fact that there are clearly too few pests in your flowerbeds to sustain such a quantity of helpers.

Population explosions

Every species needs plentiful food to breed: the more there is to eat, the greater the number of young that can be reared and which can survive.

Let's assume that the plants in your garden and the area around it are broadly the same year on year, so theoretically it would follow that the pest population should remain relatively constant. Yet the equilibrium is regularly tipped with surges in the population of aphids, butterflies and voles, and there are many reasons for this.

The first reason is good weather in winter and spring. Many animals, including humans, share a preference for dry, warm weather; the best conditions for feeling fit and healthy. The second reason is the availability of food: as we have seen, bountiful quantities mean plenty of offspring. The third and final reason is the absence of pathogens. The animal population is always at its lowest ebb in spring, right after the icy temperatures of winter. Many individual animals will have either starved, frozen or fallen prey to predators. These losses aren't replaced at this stage as offspring are not typically born in winter. In addition, many species spend the winter in hibernation or are still at the egg or larval stage. Disease can only spread when animals are on the move, coming into contact with other members of the same species. At the start of spring, movement and contact are both minimal, which means that the next generation is exposed to a reduced risk of disease.

After winter, each species attempts to regain its former strength by recouping its numbers as quickly as possible, and they have various strategies to achieve this.

Aphids, for example, reproduce asexually in spring, with female aphids giving birth to live young without the need for fertilisation: a much more efficient solution than entering an elaborate romance with a member of the opposite sex. Each female is capable of producing up to six new offspring per day, depending on the amount of food available. If the supply of juicy vegetables and rose leaves is maintained, the aphid population can keep growing at an uncompromising pace.

For butterfly caterpillars, the previous autumn is the critical factor. If female butterflies are able to lay eggs on host plants in sufficient numbers, there will be a veritable armada of caterpillars in spring. These little plant munchers will only fully develop if the weather stays mild and dry, but if they do make it, plants, shrubs and trees will be demolished until every last leaf is gone.

Voles never cease to amaze me. During the thaw after a snowy period, you can see that the chilly conditions didn't stop them from tunnelling beneath the grass, nibbling away at the roots. As the snow melts, the water gushes out of the vole holes with such force that it seems incredible that these small rodents don't freeze to death. Most, however, seem to make it successfully to spring unharmed.

The vole most typically found in our gardens here in Germany is the common vole. A close relative of the extremely unpopular European water vole, the common vole is also quite partial to a delicious meal of succulent greens. Common voles breed throughout the year, and with a gestation period of just four weeks, this has a significant impact. The young voles reach sexual maturity after just two weeks and again give birth to young of their own every four weeks. To make sure that the baby voles have a sufficient supply of milk, the mothers take turns feeding them, and the horde of voles therefore grows at a relentless rate. In good years, there can be up to two voles for every 10 m² of garden (you can work out for yourself how many might be burrowing away in your own back garden.)

The peak 'year of the vole' occurs every three years on average, but as with aphids, the key to their success is favourable weather combined with a generous supply of food.

Population explosions of aphids, caterpillars and voles are generally followed by much wailing and gnashing of teeth – and I'm not just talking about gardeners. Then comes the time when these uninvited guests really meet their match, for by the end of summer, their predators will have regained their former numbers. But although a stronger predator population does make a small contribution to population collapse, the real reason can only be seen under the microscope. The cause here is the pathogens that sweep through the 'conurbation' of caterpillars and aphids. In an insect metropolis, individuals live in close proximity, making it easy for viruses and bacteria to jump blithely from animal to animal, quickly infecting the entire population. The population subsequently collapses, sometimes to the point where there are too few individuals left for the disease to be transmitted any further. This is why a peak year is often followed by a scarce year, in a classic boom and bust cycle.

Other factors can also cause the accelerated downfall of these unpopular plant munchers. A food shortage is one that can strike a particularly serious blow to caterpillars. If green oak moth caterpillars manage to devour all of the available leaves (and an infestation can indeed defoliate an entire oak tree) before they pupate, then the time's up. If they haven't yet transitioned to the chrysalis stage, from which they'll later emerge as butterflies, their luck has run out and they'll return to humus once more.

This is also the time a number of hungry predators show up. Thanks to the plentiful food on offer over the summer, ladybirds, tits and hawks will also have had a chance to proliferate. Their offspring cause the prey to decline further, until suddenly, all is calm in your vegetable patch. The flipside, though, is that the predators will ultimately meet the same fate as their prey, albeit after a short delay: their numbers will also undergo a sharp decline. Many young animals starve or migrate, and the populations of animals we view as helpful to our gardens will return to a low plateau.

It's very difficult to stop such population fluctuations, as the swelling in numbers will have subsided naturally by the time any measures you've taken can have an effect. Many insect infestations will have passed by June, meaning they often don't last long enough to be of any benefit to tits and other birds. You can weed out or rinse off individual plants that have been accosted by these insects; however, it isn't possible to protect everything. Voles too can prove unstoppable, so the only thing I can advise here is patience. As discussed in the previous chapters, the more diverse your garden is, the more predators will be waiting in the wings to take advantage of the prey on offer.

I would also like to say a word or two about ultrasound vole deterrents. They are truly best avoided. Their effectiveness is limited, yet the acoustic pollution they cause has a significant impact on other animals. They emit a constant noise that, although inaudible to us, creates a disturbance for many species, including bats. Furthermore, bats prey on insects such as winter moths, so this is one night hunter you certainly don't want to drive away.

Birds in winter

Last autumn, I did something I'd always considered taboo: I knocked together a bird feeder and put it out in the garden. Yet, for 20 years I'd been resolutely against such an enterprise, for many good reasons. But to give you a better understanding of how I came to change my mind, I would like to contrast my previous convictions with my more recent, more balanced, arguments in favour of bird feeders.

My first argument against bird feeders was evolution. Every species is engaged in a constant battle to maintain its position in the natural world. This battle is how animals adapt to new conditions and remain genetically fit. The cold and the dramatic food shortages that accompany winter are important to the process of natural selection: only the best will survive to breed next spring. By feeding the birds in winter, however, we intervene in a way that wrenches control away from nature. For it's not just the healthy birds that help themselves to the oat flakes, fat balls and sunflower seeds; so do the weak and infirm. Wouldn't it be kinder to leave these birds to their fate?

On the other hand, our gardens and the modern landscape no longer have much in common with the natural world as it was. Where once vast stretches of land were littered with seed heads, or with endless quantities of dead wood offering shelter to billions of insect larvae, today these have mostly been replaced by swathes of tidy, barren land. If our actions, our impulse for order, causes birds to go hungry, isn't it reasonable for us to intervene and feed them? If that isn't good enough a reason, then what about good old pity? Trying to come to nature's rescue might seem slightly dubious from an ecological point of view; however, the truth remains that food is a matter of life and death for these birds, and feeding them is a good way for me and my family to express our empathy towards these vulnerable creatures.

Another argument against feeding is population dynamics. As we have seen, it's absolutely normal for up to 80 per cent of the young animals born each year not to survive to the following summer, and it's not unusual either for birds to raise two, sometimes three, broods each year to compensate for these losses. By feeding them, we help a number of young birds through the winter, leading to a larger springtime population crammed into smaller breeding territories. Feeding also causes the species composition to become skewed. Fat balls and bird food mixes really only benefit a very narrow spectrum of native bird species: these then become more widespread, often at the expense of others.

It's not easy to refute this argument, but neither can I say how much weight should actually be ascribed to it, since very little research has been conducted into the impact of feeding wild birds. The fact of the matter is this: the species composition in our cultural landscape has been artificially skewed for a very long time. Centuries ago, there was a great shift from woodland birds to grassland birds. It's difficult to say whether the modern practice of feeding garden visitors is encouraging this development or not, since it isn't only synanthropic species, such as sparrows, which show up at our bird feeders; so do woodland birds such as woodpeckers. One group of birds is certainly placed at a disadvantage by feeding, and that is migratory birds. While these birds are pottering about down south, an above-average number of young resident birds (the ones that stay with us all year round) make it through the winter. Before the transient species begin their mass exodus back to the north, most of the free territory has been taken over by those who stayed at home. When the weary travellers return, there's greater competition for food and nesting sites than there was before they left. You may be tempted to think that the solution is to hang up nest boxes to address their housing crisis, but as we'll see on the next page, this too can cause problems.

There's one big advantage to winter feeding, but it's a benefit to us humans more than to the animals: it gives us the opportunity to see species normally too shy to come out into the open. When I overcame my objection to bird tables last autumn, I had quite a surprise. Just after I started putting feed out, what should show up but some middle spotted woodpeckers: a sign of ancient, unspoiled woodland. Their appearance, right before my eyes, was tangible evidence that the ecological forest management approaches I'd been practising for 15 years were finally being reflected in the wildlife. If I hadn't put the food out, it's very possible that I never would have realised that these rare creatures had settled in the woodlands that I manage.

My conclusion is that the negative consequences of feeding birds in winter are outweighed by the positive: the joy of watching them, the knowledge you gain about which species are living locally and the act of helping birds survive in our impoverished, overly cultivated landscape. I stop putting out food from springtime so as not to skew the competition in favour of either resident or migratory birds. I also try to offer them natural sources of food by leaving a fallen tree as a refuge for insects and a source of food for woodpeckers. Our newly planted roses are old varieties which produce nutritious rosehips.

Nest boxes

If you're a gardener and a nature lover, you'll probably ask yourself at some point whether you should hang up a nest box. We have a tit box in our own garden, but nuthatches also regularly use it to raise their young. You probably already know what I'm going to say next: yes, nest boxes give rise to the same problems as bird feed. If you're simply hanging up the nest boxes for your own enjoyment, because they offer an excellent way to observe the business of rearing chicks, then go for it. Placed in a good spot, a nest box is indeed a perfect way to see the birds come and go. And there's nothing wrong with doing it for your own benefit either: an appreciation of nature is in itself a valid reason. If, however, your main reason for installing a nest box is to help the birds themselves, then it might be worth considering things from a different perspective.

First of all, there's the issue of biodiversity. In general, the nest boxes you see in garden centres and mail-order catalogues are only attractive to certain species. The diameter of the entrance hole and the size of the nest box determine which birds are able to manoeuvre themselves into this prefab housing stock. Garden owners gauge their investment according to the number of broods raised in it, and the greater the demand for nesting sites, the more likely the nest boxes are to be successful. Who, after all, buys a nest box only for it to stay empty year after year, alone and abandoned, waiting for a bird that's on the brink of extinction? When one seems not to be working, accusations are frequently levelled at its construction, and the box is written off as a poor investment. Garden centres want to keep their customers happy, and so in general, they only offer nest boxes for the most common of our feathered friends. Now, don't get me wrong: tits, nuthatches, swallows and wrens are wonderful birds, but they're not exactly rare. By providing them with a suitable nesting site, you're helping to increase an already buoyant local population. And because gardens are generally already split between several different species, with every ecological niche already occupied, supporting certain species means you shift the environmental balance. If thousands of tits or nuthatches take up residence in these artificial nests, they may raise far too many offspring, who will then storm your garden in search of insects.

Nevertheless, despite their downsides, the use of nest boxes can be justified. They let us play a part in nature and stay connected to it, and they help us maintain our sensitivity towards the environment. If you don't want to miss this chance to observe birds, yet want to disrupt the natural equilibrium as little as possible, you can take the following steps.

Note which types of birds come to your garden in summer. In most cases, your visitors will be various tits and finches. For the greatest chance of success, provide the birds with a tit box with an entrance hole diameter of 32 mm: it's this measurement and not the size of the nest box that determines which birds can take up residence inside. If you have sparrows in your garden, they will only be able to get inside a box with a 36 mm diameter entrance hole.

It's also worth having a universal nest box as then rarer species will also get a look in. What that might be depends entirely on your location. Nest boxes all have one thing in common: they emulate a hole in a tree trunk. Besides tits and nuthatches, several birds live in and on tree trunks, including small ones like treecreepers and larger species such as woodpeckers, stock doves, owls and even geese.

In nature, a hollow tree trunk is most likely to be found in a forest or a fruit tree orchard, and your local area will have a decisive impact on the success of your artificial tree hollow. If you and your neighbours have plenty of trees in your gardens, if you have a park with many trees nearby or if your property is on the edge of a wood, you might get visits from woodpeckers and other woodland birds. Why not offer these rarer birds somewhere to raise their young? There's no sense in providing more than one nest box for these larger species, though: their territories are expansive, so only one breeding pair will be able to establish itself on your patch. If the box has to be one size smaller, then a box for redstarts is perfect. The common redstart is critically endangered but is glad to make itself at home in a small garden with plenty of trees.

If, on the other hand, trees are scarce in your neighbourhood, you're better off choosing a nest box that is more attractive to grassland birds, such as barn swallows or house sparrows.

Undesirable squatters

Some species prefer nothing better than to make themselves at home in the four walls we've set aside for ourselves. What we see as brazen cheek is entirely normal for our animal compatriots; to them, there is no distinction between the natural and the artificially constructed world. Animals see houses as unusually symmetrical rock faces, filled with nooks and crannies perfect for nesting. And that's not all: these cliff-face cavities also feature mysterious thermal sources that make them a drier – and significantly warmer – option than other potential nesting spots. What a cosy place to live: no wonder some animals feel instinctively drawn to them!

Provided they restrict their visit to the attic, certain lodgers are tolerable: some bat species like to spend the summer in an attic and to raise their young there, sheltered from inclement weather. There's one disadvantage, however: bats leave their droppings indiscriminately, soiling any items you may have stored up there, as well as the floor. Your things are easily protected with a plastic sheet, though, and the payoff is that you get a privileged view of the bats as they leave the roost in the evening.

Visits from other mammals can be much less enjoyable: the pitter-patter of tiny mouse, dormouse or marten feet can make for many sleepless nights. In these cases, a live trap can be the only thing to move these uninvited guests on to a new home (preferably the edge of a wood!). To ensure your attic isn't taken over by a new gang of chancers the minute the others leave, you should investigate how they got in in the first place. In our house, the culprit was some ivy that had crept up to the height of the roof, which wood mice simply climbed to reach the rafters and the attic. Once we had cut the ivy back, we managed to nip the problem in the bud.

Martens usually enter a house through an air vent. However, don't be tempted to clog yours up with insulating foam as a friend of ours recently did. Your house will no longer be ventilated, for one thing, and, for another, martens can rip through this trifling obstacle with their sharp claws in a matter of minutes. Instead, you could stuff chicken wire into gaps or nail a piece over the vent without restricting ventilation.

As desirable as they are for nesting, our 'artificial rocks' do not, however, make suitable hibernation sites. They may be frost-free like any natural cave, but their ambient temperature is far too balmy. This summer-like warmth stimulates circulation and metabolism at such a rate that a hibernating animal would use up its fat reserves in next to no time and starve. So if you come across any lost ladybirds or green lacewings indoors, it's best to take them back outside.

Animal invaders

I have two apple trees in the garden, given to me by my parents 20 years ago when we first moved into the forester's lodge. They struggled away in the barren soil until, five years ago, I started to feed them by putting compost around their bases. They gratefully accepted the fertiliser and their growth increased considerably until last spring, for the very first time, they were in full bloom. The apples got larger and I was really looking forward to the autumn harvest. A branch (laden with apples!) was brought down by a squall of wind, and I dragged it back to the house. Upon closer inspection, I discovered that the branch was hollow inside and full of feeding marks. I opened it up and out tumbled a yellowy-white larva, around six centimetres long. This made me worried, as by now I had heard countless stories about introduced longhorn beetles, such as citrus longhorn beetles and apple borers. Was this an apple borer larva?

It was a very real possibility, and online searches did nothing to soothe my fear: this beetle first arrived in Germany in the summer of 2008 and has even made its way as far as the island of Fehmarn, in the Baltic Sea. The beetle larvae tunnel beneath the bark of a tree trunk or branch, eating and growing along the way, before pupating and emerging to the outside world in July via boreholes a centimetre in diameter. Afflicted trees were cleared and burned, and trees and shrubs up to a radius of several kilometres – anything that was a potential habitat for the beetles, such as fruit trees, rowan and hawthorn – were dowsed in insecticide. Just the prospect of this happening in our organic garden sent a shiver down my spine. Then I took a closer look. Longhorn beetle larvae – this fearsome invader – have quite a flat body which widens at the head with a mouth and pincers. My discovery, however, looked more like a butterfly caterpillar, and its body was covered with small black dots. A glance in my field guide had me breathing a sigh of relief: this was the larva of the leopard moth, a native species. Like apple borers, this moth also attacks and causes damage to deciduous trees, but since there's often just one larva in each tree, the damage stays within reasonable limits.

A visit from a certain species of ladybird can be a much greater cause for concern. Usually, the ladybird is one of the few insects that elicits a certain sympathy in us, not just because of its cheerful polka-dot wing covers; its larvae are also highly effective aphid killers. This idyll is at threat now though, with a new ladybird muscling in on the scene and making the headlines. This new arrival is the Asian or harlequin ladybird and it's currently sweeping through our gardens. Unlike longhorn beetles, the harlequin was deliberately introduced as a form of biological pest control. Organic farms in France and Belgium wanted to use these little helpers as a substitute for insecticide: a laudable aim. As is the case with our own native ladybirds, the harlequin's favourite food is aphids and one of the beetles can get through around 200 in a day. Surely, you may ask, this is a cause for celebration? If only things were so simple. Firstly, these intruders also feed on other insects, and secondly, our long-established native ladybirds were getting on perfectly well managing the aphid stocks before they came along. Harlequins reproduce extremely quickly and if food becomes scarce, they'll resort to eating other ladybirds and their larvae. So is time running out for native ladybirds? Yes, you've guessed it. Scientists fear that some species may soon disappear altogether.

These newcomers also torment humans: in autumn, they can gather in their thousands on house walls and roofs, trying to find a way inside. While they're not harming anyone or anything by doing this – they're considered nothing more than a nuisance – winemakers have real reason to fear these invaders. At harvest time, they swarm over grapes and if they end up falling into the wine must with the fruit, they secrete a liquid from their glands in panic, ruining the flavour of entire crops.

Keep an eye out for the Asian harlequin ladybird's arrival in your garden. It's relatively large (six to eight millimetres) and can have up to 19 spots. It has a black M-shaped mark on a white background on the pronotum (or a W depending on which way you're looking at it). If the ladybird has any of these features, chances are that it's one of our new residents. The ladybird can't be identified with total certainty, however, as its colouring and features can vary widely. You sometimes hear that native species of ladybird have only seven spots, but this is wrong: some European species can have over 20 and some rare varieties have the M shape too. Having said this, if the ladybird you're looking at has all three features, i.e. is large, has 19 spots and an M, the likelihood of it being a harlequin is very high.

Other migrants have made themselves at home in our climes. The Eurasian collared dove, for instance, has been moving gradually since the 1940s from South-Eastern Europe to Central Europe, slowly making its way onwards to the north-west. This beige and grey bird has a characteristic black neck collar, making it easy to distinguish from native doves and pigeons. Although slightly late to the party, the Eurasian collared dove is a typical synanthrope as it can only survive in cultivated steppe; it would have stood little chance at all in Europe's primeval deciduous forests.

The fieldfare is a bird that's followed a similar course, becoming considerably more prevalent in recent decades. Unlike the Eurasian collared dove, however, the fieldfare originally came from Eastern Europe and the Siberian taiga, moving steadily west. In many places, it's no longer a wintering bird and stays to raise its broods in the same location, where its eggs and young risk falling prey to squirrels or cranes. The parent birds' characteristic style of protecting their young from predators is likely to be the only way you'll know that fieldfare are brooding: they fly straight at the enemy, sounding their chattering alarm call.

This is all by way of introduction to some of our most recent arrivals. Some were deliberately introduced by humans and have spread intractably across their new safe haven. The other group has travelled here of its own accord and has found our agricultural landscape to be not unlike its original habitat (the steppe). Species composition changes constantly, even without the intervention of humans, thanks to the fluctuating climate. The climate has always shifted from cold to warm and back again, and will continue to do so. Habitats change as temperatures change, which is precisely why the warmth-loving beech tree has been moving northwards over the past 5,000 years. The beech's prevalence now threatens the oak, which is shuffling eastwards across Europe. Following in the wake of these trees is a broad range of animals (in the case of the beech, around 6,000 different species) that are forced to travel with them to maintain their way of life. Nature never remains static: it's in a permanent state of flux, and that's without taking into consideration the pressure humans are now placing on the climate.

Animals wild and tame

Stories about rampaging wild boar are continually popping up in the news in Germany. From front gardens and vineyards to central Berlin's Alexanderplatz, these bristly beasts are appearing everywhere and seem to have lost their fear of people. This encroachment into urban spaces is unprecedented. Now all frustrated gardeners can do is watch as their lovingly planted tulip bulbs are churned up and their lawns are trashed so thoroughly that they need to be completely reseeded.

Although wild boar like fruit and vegetables, their favourite food is meat, and they find this under the turf in the form of earthworms and voles. But why are they constantly venturing into residential areas? The primary reason is a rapidly rising population and the need to search for new habitats. According to officials such as forestry authorities and hunters, this population surge has two causes: the warmer weather brought by climate change, and greater access to food thanks to abundant beechnut and acorn crops, as well as increased maize cultivation. With all due respect, however, these explanations are quite simply nonsense.

It's true that the weather has got warmer over the last couple of decades. Warmer: there's something comforting about that word. Nobody enjoys being cold, after all – and why should wild boar be any different? Higher temperatures in winter generally mean that it doesn't freeze over. But cold is still cold regardless of whether it's plus five or minus five; it doesn't make much difference to these hardy animals. What does make a difference is that instead of snow, we're getting more rain in Germany. Damp cold is the worst and not just for humans: piglets sicken more easily and death rates increase. Climate change, therefore, doesn't benefit boar in the slightest.

This then leaves the question of whether the boar are getting more food because of agriculture and natural events. For all species, more food means that more babies survive, so this argument seems as though it could be valid. Upon closer analysis, however, it transpires that the troughs are only full for a few months each year. In the open country, pickings are rich from late summer to harvest, after which it's entirely devoid of food. In oak and beech woods, there's a particular abundance of fruit only every three to five years. These years are called mast years because in centuries past, domestic pigs were sent into the forest to gorge on acorns and beechnuts before being slaughtered in winter for truly flavoursome bacon.

By the end of December, the fruit from the trees has been polished off – another food source exhausted. At this point, the boar have to rely on their winter fat reserves. By spring, however, these reserves are depleted and there's still nothing for them to eat in the fields. Furthermore, acorns and beechnuts aren't available every year. Climate, agriculture and forest fruits, therefore, seem unlikely to have had such an influence on the consistently high wild boar populations. The chief reason, I believe, for their increase in numbers lies in the behaviour of hunters. Although hunters are among those loudly bemoaning the growing boar population, they feed the animals in secret locations in the woods all year round so they have enough animals to hunt. Every boar slain is fed an average of 130 kilogrammes of grain maize. With quantities such as this, the animals might as well be fattened up in pens. The only measure that would effectively quell this swelling tide of boar would be a ban on feeding them, but unfortunately initiatives pushing for this never get very far because of a strong hunting lobby, which has influence in all state parliaments across Germany. As a garden owner, your only option is to protect your land by erecting a huge metal fence set deep into the ground.

Strong population growth is one reason why more and more foxes, deer and wild boar are moving into our cities, but there is another: the Serengeti effect. Have you ever been on safari in Africa or seen one on TV? What's noticeable about the animals in Africa's national parks is their familiarity with humans and their total disinterest in them. You can drive a jeep right up to the lions, elephants, zebras and gazelles, as close as a few metres away, without disturbing them in the slightest. Just outside the boundaries of the park, however, these idyllic scenes are no more. Outside the park, you're likely to see animals much less frequently and the reason for that is simple: here they're hunted, both legally and illegally, and they've therefore come to fear humans. The exact same thing is seen in Central Europe: while out hiking, the chances are low that you'll come across a red deer or roe deer. And that's with around 50 to 100 animals per square kilometre in many regions, so on a hike averaging 15 kilometres, you'd think you'd have a reasonable chance of seeing something. The reality is, of course, very different and this is purely down to hunting. Our wild animals are permanently stalked by hunters (in Germany alone, 350,000 hunting licences have been issued), meaning that they live in a constant state of fear and anxiety. They moderate their behaviour according to that of us humans. Although deer need to roam widely and graze all day to get their fill of grassy vegetation, they only emerge from the woods and hedges at night. They know from bitter experience that they may be shot at any time up until dusk. The gunfire stops once total darkness has fallen, and only then can they eat in the meadows undisturbed. By day, the animals flee to wooded areas where they can't be seen and where their grumbling stomachs lead them to nibble on buds, tree leaves and sometimes even bark.

Now let's turn our attention back to your garden. Since hunting is banned in residential areas, your garden is like a miniature national park. One individual garden is, however, too small for wild animals to realise it's a safe zone. Several plots of land together, or even an entire street of gardens, is a different matter entirely. Once foxes cotton on to the fact that neither you nor your neighbours pose any danger, their behaviour will change completely. Just like in the Serengeti, they'll become active during the day and will return to a more normal rhythm of life. It will take them some time to become as tame as the animals in a national park; our gardens are clearly too small to provide a hunt-free oasis for the local wild animal population. And if just one animal has a bad experience, its nervousness will spread to other animals of the same species, and that trust will be lost. Nevertheless, even if they visit occasionally, you'll be granted a rare glimpse of them during the day.

To attract wild visitors, some garden owners leave out a regular supply of apples or grain. We've already discussed the reasons why we might decide to feed wild birds, but in principle, we shouldn't encourage wild animals to become too accustomed to humans. Too much familiarity can have a negative impact on both parties. There are dangers in our midst that wild animals are unable to adapt to quickly enough. They realise too late that cars, lawnmowers and prowling pets pose a threat to them, and trusting squirrels, birds and deer pay for their curiosity with their lives.

Vice versa, the animals we feed can even begin to cause a nuisance to us. My previous neighbours, for instance, fed squirrels with peanuts and they soon became tame. Every day, these little reddish-brown scamps would appear at the patio door. If nobody was in at the usual time or if my neighbours didn't respond quickly enough, their impatient guests would scratch at the door with their front paws. The racket soon became a real annoyance, not to mention the damage to the door frame. We had been tempted to leave out food for these cute little visitors too, but thankfully we learnt from our neighbours' bad experience.

On the question of taming and keeping wild animals, I'm constantly torn. On the one hand, we humans interfere with the natural world quite enough without doing it in our free time as well. But on the other hand, we only truly protect what we love. And what better way is there to foster a love of animals than by taking them under our wing and looking after them? I think it's better to keep a wild animal as a pet full-time than to lure wild animals to our houses by feeding them occasionally, the difference being that a pet has no influence on the natural world, stays with humans the whole time, and gives people a fuller sense of what it means to look after and live with an animal. The wild animals we feed, however, adapt their behaviour to such an extent that they become less capable of surviving on their own, having grown dependent on the welfare provided by humans.

What would be the problem with keeping a tame, highly intelligent crow or jay? At the moment, the only thing standing in the way of this is legislation, while ironically in Germany you're allowed to shoot them. Critically endangered exotic birds, fish and lizards, by contrast, can be trapped and kept as pets, and although they require a licence, the capture of animals is all too often carried out illegally in their tropical homelands.

Abandoned offspring

It's a situation most gardeners will experience at one time or another: you go out to pick some lettuce or do a spot of raking, and, all of a sudden, you see two anxious little eyes staring up at you. Perhaps it's a young bird that has fallen from its nest. We take pity on such vulnerable creatures and instinctively want to do something to help them. Doing so could pose a danger to these young birds, however, and it's difficult to tell if they actually need our help or not.

A few months ago, a local man brought me a young buzzard that he'd found squatting, apparently helpless, on the woodland floor. Towering in the tree above it was the eyrie and there were the buzzard's siblings, who were calling for their parents. The young bird already had its adult plumage and so was close to being fully independent.

The bird's finder was very disappointed when I told him to put the buzzard back under the tree where he had found it. I'm sure he thought I was a heartless brute! Well, regardless of his opinion of me, this action saved the bird. After a few days, I went to check on how the foundling was doing. There it was, perched on a tree stump, pulling at a piece of meat and bones that had most likely been brought to it by its parents. After finishing its meal, it flapped its wings a little before taking to the air and flying to the top of the neighbouring tree.

Another case involved a fawn. The young deer had been found in a field by some children. They took the animal back home, at which point their mother called me. She was completely at a loss as to what to do with it and asked if I would mind coming to pick it up. The children couldn't remember exactly where they had found it, so taking it back was no longer an option.

I was a little annoyed as I expected the rural population at least to know by now that you should never move a fawn, even if it looks like it's abandoned. When the fawn is new born, the doe will leave it alone at the edge of the wood to go and forage for food. From time to time she'll come back to check on it and give it a drink of milk before returning to her search. Once the young deer can walk, it'll follow its mum wherever she goes.

So here I was with a small, trembling baby deer. All attempts to feed it from a bottle failed and the little one starved.

In my youth, I reared a considerable number of birds, rabbits and later martens. Sometimes I was able to save them and sometimes I wasn't, which is why I always think twice before interfering with abandoned baby animals. Here are a few simple rules to consider before you decide to take a young animal into the warmth of your home.

The following rule applies to birds: the larger the chick, the less help it needs. If the young bird's adult plumage is already showing and it can walk well, its parents are very likely still feeding it, no matter how far it wanders from the nest while exploring your garden. If, however, the bird is still covered in down or practically bald, it won't survive on the ground. The first thing you can do to help is to look for the nest. You may be able to just pop the young bird back inside. Don't worry about doing this: your scent won't stop the parents from feeding it. If you aren't able to place the chick back in the nest, you can either foster it yourself or hand it over to an organisation that specialises in the care of wild birds. In the UK, the RSPB or the RSPCA are the best places to turn for advice, and both organisations publish good information online.

Hedgehogs are a particular focus of attention for many animal protection campaigners, and their relatively large size and wild yet innocuous nature makes them especially symbolic of our connection to nature. There are so many things to consider when feeding hedgehogs that if you suspect you've found an underweight or young, abandoned hedgehog, I recommend that you feed them under the guidance of a specialist. Contact a vet or conservation group in your area for advice or seek out specialist literature.

When in doubt, one simple, if rather radical rule, applies to mammals: leave the animal exactly where you found it. The baby's mother is usually nearby, is looking after it, and is teaching it how to become independent. You should only attempt to help if the baby clearly looks underfed or if there's still no sign of the mother after several hours. In these instances, though, the outcome is often disappointing as the abandoned babies are usually sick and will soon die.

One other important aspect is worth bearing in mind. Animals produce a lot of offspring because nature is merciless in sorting out the weak from the strong, leaving only the healthy to survive. By feeding up the losers, we're potentially weakening the species' local population. It may sound harsh, but providing help is often counterproductive.

There are only a very few cases when intervention is recommended: when the mother animal has had an accident, can't be reached for some reason, or if the young animal has been injured (and it goes without saying that you should also help injured adult animals).

If you want to help baby animals, the best way of doing so is to have a natural garden with ecological niches; that is, pockets of wilderness. By avoiding the use of chemicals and leaving some areas uncut or untended, you will give many young animals the best chance of surviving their first year.

11

EXPERIENCING NATURE
WITH ALL OUR SENSES

THE natural phenomena we've considered in this book are just a few examples of what can be experienced in your garden. There are thousands of others besides; all you need to do is to take notice of them. Think of the outraged reaction of songbirds as they catch sight of a raptor, for example, or how the smell of the air changes when a rain front is approaching. But more important than being able to recognise these phenomena is honing your senses to appreciate the diversity of your garden; so let's take a closer look at the sensory tools at our disposal.

Night vision

Humans are visual animals: our sense of sight is more highly developed than our sense of hearing or smell. As we were once primitive creatures of the wide, open steppe, this made sense as sight can extend kilometres into the distance, while hearing and smell are more localised. This meant that both enemies and potential food could be spotted in good time.

Long-distance vision is in our genes as human beings, and our landscape has been designed accordingly. The dark forests and trees that once obstructed our field of view have been replaced by endless open grassland. Our fields and meadows follow this archetypal model. It's only the species composition that's changed, to the wheat, maize and barley we grow now. This desire for a grassland microcosm is even reflected in our gardens. You may think that using hedgerows and fences as screens contradicts this, but the point of these barriers is not to limit our view but to get some privacy from our neighbours.

The eye is only able to perceive a narrow band of electromagnetic waves, which we refer to as 'light'. Every time it gets dark, it becomes abundantly clear just how much we depend on our sense of sight. Our ability to determine colour disappears as soon as twilight falls; as they say, at night all cats are grey. When the light level drops below 0.1 lux, we can hardly see anything at all; meanwhile, on a sunny day, light has a strength of 100,000 lux.

When it's dark, the only thing that's missing is light. Information reaches all of our other senses in the form of sound, smell and touch just as before. Yet, although nothing has changed in these respects, the night landscape seems like a completely different world. If we're on our own and something rustles in a bush, most of us feel a slight sense of unease. This shows us just how dominant our sense of sight is and how frightening it can be when we are deprived of it.

But even when we can see just fine, it can still be too dark for various life forms, including us humans. This is often the case indoors during winter.

A clear warning sign is when a houseplant produces unusually long shoots and its leaves turn yellow. This means that the room isn't light enough – and this can affect your health too. If the level of light remains below 2500 lux (the amount of light in the garden on a dull winter's day) for an extended period of time, it can lead to seasonal affective disorder. Being in a badly-lit room is like being in a permanent state of winter. To prevent this, you should make sure that the rooms where you spend your days are sufficiently well lit and that you go outside regularly, even if the weather doesn't make it a particularly pleasant proposition.

While we're on the subject of lighting, our love of light is the reason why, in built-up areas, we've turned the night into day. We light up our houses, which can indeed be beneficial to our health in winter, while outside on the streets, we defy the rule that the night is a time of darkness. Our preference for flooding our neighbourhoods with light not only consumes a lot of energy (three to four billion kilowatt hours of energy are used in Germany alone each year), it also poses a problem of an entirely different order to the environment. Our electrification of the night is a form of air pollution. You can discern this for yourself: after your eyes have become acclimatised to the light levels, you should be able to see the Milky Way on a clear night. These days it can only be seen in the countryside, because there is a constant haze of exhaust fumes and condensation suspended in city air. The light from streetlamps shines through this artificial fog, spreading a diffuse glow over our urban and suburban areas at all times. This stronger light swallows up the gentle shimmer of the Milky Way and the faint light of the other stars. In the countryside, almost 3,000 stars can be seen with the naked eye, but in the towns and cities this is reduced to fewer than 1,000. Although this might not constitute an environmental issue, it does rob us of our enjoyment of an aspect of the natural world.

For some animals, on the other hand, streetlamps and garden lighting can have fatal consequences. Moths, for example, use celestial bodies to orientate themselves, navigating by flying at a constant angle to the Moon. The Moon's great distance from the Earth means that, as the moth travels forwards, it seems to stay at a constant angle relative to the moth's flight path, making night flight a piece of cake. Or at least it should do. For these little pilots, our artificial lights shine as brightly as the moon, but with one vital difference: they're much closer. When a moth flies past a light bulb – a mysterious, false moon – it finds that suddenly the light is behind it rather than in front of it. The insect then believes its flight path to be crooked rather than straight. The moth changes its direction so that it can continue flying parallel to the 'moon', but instead follows a circular course around the light source, culminating with the moth crashing into the lamp. When this happens, there is no escape; inexplicably for the moth, no matter where it flies, the 'moon' is always behind it. If this state of confusion continues too long, the animal will die of exhaustion. In some places, predators have adapted to the situation. On warm summer nights, for instance, bats can be seen patrolling the street lamps, where moths and other confused flying insects are easy prey as they circle helplessly around the lightbulb.

For this reason, you should close your blinds or curtains as soon as you switch on your lights after dark to avoid bringing the street-light spectacle – complete with bats – to your living room window. The same applies to all-night garden lighting. It may look romantic, but wouldn't it be better to come to a compromise and keep it on for just a few hours? Security lighting on pavements could also be switched off for part of the night at least. Once the residents have gone to sleep, shouldn't we let darkness fall? Likewise, solar powered lighting, which runs all night long, wouldn't be my first choice.

Researchers at the University of Exeter have found that night lighting changes the species composition on the ground around the light – and this change may be permanent. They discovered that small predators and scavengers, such as spiders and woodlice, gather in larger than normal numbers beneath street lamps, even during the day. Further research will be required to determine what implications this has for the ecosystem.

At twilight towards the end of June or in early July, a very special spectacle can be seen in the gardens of Central Europe: fireflies. These bioluminescent beetles produce light from their abdomens to attract a partner. The lights seen flying through the air belong to the male fireflies who take flight in search of females. These females also illuminate but sit on the ground as they're unable to fly, which makes it easy to tell the difference between the two sexes, even at night. Two different species of firefly are responsible for this impressive sight: the Central European firefly (*Lamprohiza splendidula*) and the common European glow-worm (*Lampyris noctiluca*).

One other bioluminescent beetle, the short-winged firefly or lesser glow worm (*Phosphaenus hemipterus*), can be observed in late summer, in the early hours after midnight. Compared with the other two you will only snatch glimpses of them in bushes or foliage but nevertheless, reducing the amount of artificial light in the garden can help them find a partner.

I've become used to only switching on the lights in the evening when I absolutely need to and because of this, I'm constantly discovering something new. To give you an example, I was once taking the dog for a walk, and as I was walking down the driveway, I heard several loud clapping sounds. Above us, I saw the dim outline of a large bird circling; in this darkness, it could only be an owl. Back in my office, I reached for my field guide straight away and discovered that these loud wing claps are made by long-eared owls during their courtship display.

So, I'd recommend paying more regular visits to your garden in the evening and leaving the torch behind. You'll soon see that it has more to offer than you might at first have realised.

Incensed by scent

We've established that humans are visual animals, but this doesn't mean that our other senses are completely useless. In the modern age, the flood of information we're exposed to naturally means that the gap between our senses is growing: the screens we rely on appeal to our eyes, after all, rather than our nose. All around us, however, are a great many different scents and smells. Recently, researchers have made a ground-breaking discovery: it turns out that plants talk to one another. This doesn't mean that plants have vocal chords, of course. No, instead they communicate by wafting different 'scent messages' through the garden. It's by no means news that plants can communicate to animals by dint of smell. The aromatic scent of flowering plants invites certain insects to stop by for a sip of nectar (and to pollinate them while they're at it). Plants use their flowers and scent to target particular species of insect. The pawpaw uses pretty purple flowers and a repulsive carrion-like stench to attract flies, while our European fruit trees prefer the attention of honeybees, releasing fragrances that are more pleasing to us. We've been aware of this method of cooperation, this style of communication, for thousands of years. What's new now, though, is the concept of plants chatting amongst themselves. Trees, for example, warn each other about insect attacks by giving off a chemical distress signal. This message prompts trees of the same species to produce defensive chemicals that are then stored in their bark. Researchers now believe that most plants communicate with others in their species.

This is a notable discovery for several reasons. Firstly, the arbitrary lines humans have drawn between plants and animals have become blurred as demonstrated by fungi. We must now concede that plants have senses and feelings, such as pain, hunger and thirst. Secondly, it's becoming clear that there are a great many natural processes that we still don't fully understand and that cannot be reconciled with our – often very basic – explanatory models.

Let's return once more to your garden. You and your nose also play a part in the garrulous conversation taking place between the shrubs and perennials. Let's consider roses for a moment. Roses are sought after for their colour, as well as their scent, so the signal they send out in a garden centre is an alluring 'Come hither!' Of course, we can also describe this in a more detached, scientific formulation: a breeder selects a certain plant variety for its marketable scent, and this is what makes that variety successful. This is the same scenario; it's just expressed in a less frivolous way. We're not used to letting our feelings influence how we describe science. But then again, why shouldn't we? By translating plant language into our own, by translating the scent into a direct request, we get much closer to the true meaning the scent is intended to convey.

Plants emit warning signals when under stress, and this also applies to the plants in your garden. If the conditions aren't right, your wards start to feel uncomfortable and this negative mood soon wafts over the grass, trees and shrubs. Conversely, when plants feel comfortable, when they're happy with their location and have enough food and water, these stress signals are absent.

Is it a coincidence that gardens like this are such a great source of relaxation? No one can really explain why, but surely, it's not inconceivable that our sense of smell may be able to alert our subconscious to recognise an intact ecosystem, one where all is well and life is good.

Smells of a completely different nature also waft through our gardens. Cats leave malodorous signals on cars, flowerpots and fence posts, warning other cats to stay out of their territory. As we have seen, many other mammals, such as martens, foxes and mice, add their own scent to this potent mix.

There are so many different smells waiting to be discovered, from the sweet, aromatic fragrance of pine trees on a warm summer's day (the needles' essential oils), the tangy aroma of oak leaves in autumn, to the damp, musty greeting sent out by ground fungi after a rain shower. If you're open to the different 'scent messages' around you, you'll get much more out of your garden than you can with your eyes alone.

Tuning in our ears

It's not just our sense of smell that is poorly developed compared to our sight: our hearing is also relatively weak. It is, of course, entirely sufficient for human communication, which is really quite loud, and we're also able to pick up a considerable amount of the noise that other creatures make. Bird song is perhaps the best example. Many bird species can only be identified by their calls, since these shy creatures prefer to hide away, out of sight in the treetops. The ones we are able to catch a glimpse of often cross our field of vision so quickly that it's impossible to get an accurate identification. What may at first look like an ordinary wood pigeon, for example, may turn out to be an extremely rare stock dove, a hole nester (their preference is for abandoned woodpecker holes). The species are the same size, grey and have the same silhouette in flight. Instead of having a white ring on its neck, however, the stock dove's neck shimmers a greeny blue.

Although these differences can rarely be seen as the bird flits between the trees, the call is a definite giveaway. The wood pigeon sings a distinctive 'coo coooo coo cu cu', while the stock dove's call is a straightforward 'coo'. I regularly hear the stock dove call during my summer walks in the forest, although I've never definitively identified one by sight.

Many other species, apart from birds, draw attention to themselves by their calls. Let's start with one of the smallest mammals, the mouse. Its shrill whistling sound is mainly heard in meadows of tall grass, and although not particularly loud, it's enough for a fox: a concerto of whistles promises a good catch.

You can tell a fox, meanwhile, by its hoarse, high-pitched bark. It sounds a little like a howl, but just two seconds long. These mouse hunters are becoming ever more common in residential areas and have even gained a foothold in the centre of our cities, so it's worth keeping an ear out for one on a quiet evening.

Over the years, you can certainly train your sense of hearing – or more precisely, your brain – to hear noises in nature. Of all the information received from our environment, our 'thinking organ' pays special attention to things of importance to us, making them seem louder. For me, it's crane calls. These large birds fly over my garden twice a year as they migrate and are, for me, a symbol of an intact ecosystem. Sometimes, their formation flies less than 100 metres above our house and when it does, I can even hear the beating of their wings.

The crane's classic call has become so etched in my mind that I can hear one, no matter how faint, from a great distance away. Last autumn, I was able to hear a flock fly past even with the window closed and TV on. Find out what your own favourite garden sound is: perhaps it's the song of a blackbird at dusk, the rustle of a hedgehog under a hedge or the brisk buzz of a bumblebee in the shrubs. There are so many amazing things to hear out there beyond the everyday tumult of cars and planes.

And in winter, when all sound has been deadened by snow, you can experience moments of complete and utter silence. For a trained ear, this is something truly special, as silence is so rare in our densely populated environment.

12

A RETURN TO NATURE

D ON'T worry, I'm not about to urge you to surrender your garden entirely to nature's whims. Ultimately, I view gardens as a good compromise that benefits both humans and the environment alike. Nevertheless, this compromise shouldn't be taken as a contract written in stone for both sides to abide by without complaint. We can't ask the natural world for its opinion, after all, so it's the gardener who sets the rules. No matter how fair these rules may be, though, nature will never stop trying to reclaim the entire space for itself.

Grass is a prime example of this. Though we maybe accepting of weeds and mosses, we still like to keep the lawn tidy and subject it to a regular mow. If you were to leave it to its own devices, you'd end up with a meadow with grass a metre tall. Gradually, more and more trees would be able to take a foothold until at last, after 100 years or so, your garden would be replaced by a wood. Every passage of the lawnmower keeps nature in its place. But still it insists on trying to get the better of you. So you want tidy, weed-free grass? You'd better think again: more often than not, that pesky moss will make a comeback. And it's hardly surprising; your actions in the garden give them all the help they need to spread. By taking away the grass cuttings, you also take away essential nutrients, causing the soil to starve as the years go on. Mosses don't ask for much – they can even make do with sheer stone – so this situation is very much to their advantage. With every batch of grass cuttings removed, the less chance moss's competitors have to grow. Now all that's needed for the moss to become unstoppable is a little moisture, all-too-willingly proffered by keen lawn cultivators and their garden sprinklers.

There are a few things you can do, however, to get rid of it. Fertilise your grass regularly with dung manure and it will become strong enough to break back through the moss. Moss can also be removed by scarifying or raking your lawn. These solutions will only help, though, if you're persistent. In this sense, it's true that the more artificial the garden, the more effort and expense are needed to maintain it. In our house, we have come to an amicable arrangement with the moss: it's allowed to grow wherever it wants and in turn, I get away with not having to do as much in terms of lawn maintenance. I simply mow the lawn and I don't bother to pick up the cuttings (and after a few days, it's all dealt with by the worms and other creepy-crawlies anyway). All of this aside, walking across a carpet of moss feels very pleasant underfoot.

But nature tries to regain control of your garden in more subtle ways, and the key here is wood. Decking and raised beds, tables and benches, fences and sheds: this wonderful natural material can be used to make them all. But the moment you place these items in your garden, an army of fungus and insects sets about trying to destroy them. These little critters don't care whether they build their nests in an overturned tree or a garden bench: wood is wood, after all. The deciding factor is moisture. Just like ordinary garden plants, fungi need a certain amount of water to survive. Fungus becomes active and starts spreading its filaments through garden furniture when the moisture of the wood reaches around 25 per cent. Roughly speaking, there are two different types of fungus: brown rot and white rot fungus, each of which feasts on different components of the wood. Wood is formed mainly from cellulose fibres and lignin in a structure rather like fibreglass. The fibres are coated in resinous lignin, which helps keep the wood cells hard and flexible.

White rot fungus is a lignin devotee and gobbles it up until white, fibrous pieces of cellulose are all that remain. Brown rot fungus works the other way around. Its favourite meal is cellulose: the leftover lignin deepens the brown colour of the wood and makes it crumble away.

There are ways of spoiling the fungi's appetite, however. The best way is simply to keep the wood dry. When moisture is below 25 per cent, the invaders are brought to a standstill. By way of comparison, the wood used to make the furniture in your home generally has a moisture content of 12 per cent. Storing the patio furniture under a shaded veranda can keep moisture down to less than 20 per cent. This method of stopping fungus is called constructive wood protection. If garden furniture ever gets wet, bringing it under cover to dry out will often be enough. If, on the other hand, the furniture is left out on the lawn for weeks, fungus will launch its attack, always starting with the feet. When the feet are standing on damp ground, absorbing water like straw, it's no wonder a fungus feels at home. And since air contains more spores than previously suspected, a fungus can establish itself within minutes. Researchers in Mainz have discovered that for every cubic metre of air, there are between 1,000 and 10,000 spores – this means up to ten spores in every breath we take. Wood is therefore often infected with fungus before it even reaches the garden. If you want to get rid of fungus without using chemicals or protective paints, you should dry it out. Do this and the wood will last for centuries.

The same applies to insects. An infestation is when they lay their eggs either in or on the wood. Their larvae then burrow their way through the wood cells, which often contain nutritious sugary residues. But besides eating, they also have to drink and if the wood is too dry, the larvae will die off before they can pupate into adult beetles, and before they can nibble holes all over your wood.

I'd like to say a thing or two about chemical wood protection: it can be tempting to paint garden furniture and just leave it outside in all weather. As the years go by, however, the stain will be weathered away and washed into the soil by torrential downpours. Once it's in the ground it can adversely affect insects and other creepy-crawlies, which is why you should either switch to natural pigments (if it's the colour that's important) or choose more durable woods. Certain woods, which include oak, larch, Douglas fir and Robinia (false acacia), have an innate resistance to weathering and last for many years, even in damp conditions. Other options include imported woods such as teak or eucalyptus, but please always look out for the FSC logo that confirms it's certified by the Forest Stewardship Council.

At some point or other, even the most durable of products will reach the end of their lifespan and turn to humus once more. In your garden, millennia-old forces are at work, life forces that are virtually untameable.

So what about us? How far removed are we from our environment and how dulled have our senses really become? We often compare humanity's performance in these areas with animals and in this comparison, our species generally comes off badly. In terms of our sight, humans and animals are pretty much on a level playing field, but throw hearing, smell and touch into the mix, and we humans are soundly beaten. We often speak admiringly about the perceptive faculties of dogs, cats and birds, but in doing so we forget that the blueprint of the human body is built around the very same principles: our senses equip us for life in the natural world, not life in an office or on a sofa at home. Yet our everyday life is dictated by artificial ecosystems, allowing us to lose sight of our biological origins.

Our brain is designed for so much more than merely working on a computer or driving a vehicle: it's the most important tool at our disposal for making sense of our environment. With the help of our old grey matter, we can sharpen our senses enough to match the sensory abilities of our fellow creatures.

I'm not advocating a return to our roots or a rejection of modern life: I like my creature comforts too much for that. No, what I'm really interested in is reclaiming our sensitivity to nature and reawakening our powers of observation which, up until now, have been buried under the clutter of modernity. When we use our senses at full capacity, we access the wealth of thrilling and calming experiences waiting for us just outside our back doors, in nature and in our gardens. The world seems to expand when we're able to appreciate it in all its diversity. I hope you find many new discoveries when you're out and about, and that, like me, you discover a world that's so much bigger than it first appears.

INDEX